Saxon Math
Course 1

Stephen Hake

Reteaching Masters

A Harcourt Achieve Imprint

www.SaxonPublishers.com
1-800-284-7019

Introduction

Monitor student progress during daily instruction and evaluate after each Cumulative Test. Assign reteaching masters as necessary to reinforce mathematical concepts and skills.

Use Individual Test Analysis Form B from the *Course Assessments* book to identify those concepts that may be causing students difficulty. Complete a separate form after every Cumulative Test for each student who scored below 80%. Not all students will have mastered a new concept the first time it is assessed. However, if a student has not mastered a new concept after repeated practice and assessment, then reteaching is indicated.

Each reteaching master contains a concise summary of the new concept introduced in a lesson or investigation, including annotated examples. The lesson summary is followed by a set of practice items to support reteaching and remediation of the new concept.

ISBN 13: 978-1-5914-1815-3

© 2007 Harcourt Achieve Inc. and Stephen Hake

All rights reserved. This book is intended for classroom use and is not for resale or distribution. Each blackline master in this book may be reproduced, with the copyright notice, without permission from the Publisher. Reproduction for an entire school or district is prohibited. No other part of this publication may be reproduced or transmitted in any form or by any means, electronic or mechanical, including photocopying, recording, taping, or any information storage and retrieval system, without permission in writing from the Publisher. Requests for permission should be mailed to: Paralegal Department, 6277 Sea Harbor Drive, Orlando, FL 32887.

Saxon is a trademark of Harcourt Achieve Inc.

Printed in the United States of America

3 4 5 6 7 8   0928   14 13 12 11 10 09

Table of Contents

Reteaching Masters

Section 1
Lesson 1 1
Lesson 2 2
Lesson 3 3
Lesson 4 4
Lesson 5 5
Lesson 6 6
Lesson 7 7
Lesson 8 8
Lesson 9 9
Lesson 10 10
Investigation 1 11

Section 2
Lesson 11 12
Lesson 12 13
Lesson 13 14
Lesson 14 15
Lesson 15 16
Lesson 16 17
Lesson 17 18
Lesson 18 19
Lesson 19 20
Lesson 20 21
Investigation 2 22

Section 3
Lesson 21 23
Lesson 22 24
Lesson 23 25
Lesson 24 26
Lesson 25 27
Lesson 26 28
Lesson 27 29
Lesson 28 30
Lesson 29 31
Lesson 30 32
Investigation 3 33

Section 4
Lesson 31 34
Lesson 32 35
Lesson 33 36
Lesson 34 37
Lesson 35 38
Lesson 36 39
Lesson 37 40
Lesson 38 41
Lesson 39 42
Lesson 40 43
Investigation 4 44

Section 5
Lesson 41 45
Lesson 42 46
Lesson 43 47
Lesson 44 48
Lesson 45 49
Lesson 46 50
Lesson 47 51
Lesson 48 52
Lesson 49 53
Lesson 50 54
Investigation 5 55

Section 6
Lesson 51 56
Lesson 52 57
Lesson 53 58
Lesson 54 59
Lesson 55 60
Lesson 56 61
Lesson 57 62
Lesson 58 63
Lesson 59 64
Lesson 60 65
Investigation 6 66

Table of Contents, continued

Section 7
Lesson 61 67
Lesson 62 68
Lesson 63 69
Lesson 64 70
Lesson 65 71
Lesson 66 72
Lesson 67 73
Lesson 68 74
Lesson 69 75
Lesson 70 76
Investigation 7 77

Section 8
Lesson 71 78
Lesson 72 79
Lesson 73 80
Lesson 74 81
Lesson 75 82
Lesson 76 83
Lesson 77 84
Lesson 78 85
Lesson 79 86
Lesson 80 87
Investigation 8 88

Section 9
Lesson 81 89
Lesson 82 90
Lesson 83 91
Lesson 84 92
Lesson 85 93
Lesson 86 94
Lesson 87 95
Lesson 88 96
Lesson 89 97
Lesson 90 98
Investigation 9 99

Section 10
Lesson 91 100
Lesson 92 101
Lesson 93 102
Lesson 94 103
Lesson 95 104
Lesson 96 105
Lesson 97 106
Lesson 98 107
Lesson 99 108
Lesson 100 109
Investigation 10 110

Section 11
Lesson 101 111
Lesson 102 112
Lesson 103 113
Lesson 104 114
Lesson 105 115
Lesson 106 116
Lesson 107 117
Lesson 108 118
Lesson 109 119
Lesson 110 120
Investigation 11 121

Section 12
Lesson 111 122
Lesson 112 123
Lesson 113 124
Lesson 114 125
Lesson 115 126
Lesson 116 127
Lesson 117 128
Lesson 118 129
Lesson 119 130
Lesson 120 131
Investigation 12 132

Name _____

Reteaching 1
Math Course 1, Lesson 1

- **Adding Whole Numbers and Money**
- **Subtracting Whole Numbers and Money**
- **Fact Families, Part 1**

- To add money, line up the decimal points.
  Then add each column starting on the right.

  Example: $\begin{array}{r} \$\ 1.25 \\ \$12.50 \\ +\ \$\ 5.00 \\ \hline \$18.75 \end{array}$

- When subtracting, put the starting amount first. Write $6 as $6.00.
  Then borrow across **all the zeros** in one step.

  Example: $\begin{array}{r} \overset{5\ 9\ 1}{\$6.00} \\ +\ \$1.23 \\ \hline \$4.77 \end{array}$

- When you learn **one** fact family, you know **four** facts.

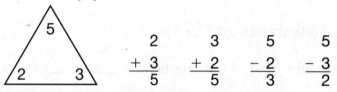

Practice:

1. What is the sum of 2518, 457, and 1263?

   _____

2. $4.90 + $0.65 + $23 =

   _____

3. 3782 − 469 =

   _____

4. $3 − 3¢ =

   _____

5. Jake paid $5 for a $3.96 sandwich. How much money should he get back?

   _____

6. Use the numbers 3, 9, and 12 to make two addition facts and two subtraction facts.

   _____

   _____

   _____

   _____

Saxon Math Course 1

Name _____

Reteaching 2

Math Course 1, Lesson 2

- **Multiplying Whole Numbers and Money**
- **Dividing Whole Numbers and Money**

Multiplication
- When multiplying by a two-digit number, multiply twice.
- When multiplying dollars and cents, the answer will have cents places.

Forms of Multiplication
$2 \times 3 \qquad 2 \cdot 3 \qquad 2(3)$

Division
- Use *short* division with one-digit divisors.
- Use *long* division with two-digit divisors.
  Remember: Put a digit above each digit. Use zero as a placeholder.

Forms of Division
$4\overline{)12} \qquad 12 \div 4 \qquad \frac{12}{4}$
"Twelve divided by four" Say the dividend first.

Practice:

1. Alicia bought 3 bracelets for $3.25 each.

   What was the total cost of the bracelets? _____

Simplify 2–5.

2. 73
   $\underline{\times\ 28}$

3. $515 \cdot 23 =$ _____

4. $9\overline{)3744}$

5. $\frac{322}{14} =$ _____

6. Use the numbers 7, 9, and 63 to make two multiplication facts and two division facts.

   _____

   _____

   _____

Name _____

Reteaching 3
Math Course 1, Lesson 3

- **Unknown Numbers in Addition**
- **Unknown Numbers in Subtraction**

The quantity on either side of an equals sign is the same.
$$4 + 5 = 9$$

A letter can take the place of an unknown number.
$$4 + x = 9$$

Addition: addend + addend = sum
To find a **missing addend, subtract** the known addend from the sum.
Examples: $2 + a = 5 \longrightarrow 5 - 2 = a \longrightarrow a = 3$ $b + 3 = 5 \longrightarrow 5 - 3 = b \longrightarrow b = 2$

Subtraction: minuend − subtrahend = difference
To find a **missing minuend, add** the difference and subtrahend.   To find a **missing subtrahend, subtract** the difference from the minuend.
Examples: $n - 3 = 2 \longrightarrow 2 + 3 = n \longrightarrow n = 5$ $5 - y = 2 \longrightarrow 5 - 2 = y \longrightarrow y = 3$

Practice:

Find each unknown number.

1. $a + 12 = 30$

    $a =$ _____

2. $m + 32 = 59$

    $m =$ _____

3. $w + 47 = 81$

    $w =$ _____

4. $8 + b = 20$

    $b =$ _____

5. $47 + p = 82$

    $p =$ _____

6. $89 + k = 125$

    $k =$ _____

7. $c - 15 = 12$

    $c =$ _____

8. $s - 26 = 15$

    $s =$ _____

9. $t - 38 = 52$

    $t =$ _____

10. $49 - d = 36$

    $d =$ _____

11. $92 - f = 67$

    $f =$ _____

12. $2000 - d = 1215$

    $d =$ _____

Reteaching 4

Math Course 1, Lesson 4

- **Unknown Numbers in Multiplication**
- **Unknown Numbers in Division**

$5w$ means "5 times w." $\frac{12}{x}$ means "12 divided by x."

Multiplication
factor × factor = product
To find an unknown **factor, divide** the product and the known factor.
Example: $5w = 20 \longrightarrow 20 \div 5 = w \longrightarrow w = 4$

Division
$\text{divisor}\overline{)\text{dividend}}^{\text{quotient}}$ $\frac{\text{dividend}}{\text{divisor}} = \text{quotient}$ $\text{dividend} \div \text{divisor} = \text{quotient}$
To find a missing **dividend, multiply** the divisor and the quotient.   To find a missing **divisor, divide** the dividend by the quotient.
Examples: $\frac{n}{3} = 6 \longrightarrow 6 \times 3 = n \longrightarrow n = 18$ $y\overline{)35}^{\,5} \longrightarrow 35 \div 5 = y \longrightarrow y = 7$

Practice:

Find each unknown number.

1. $4x = 32$

    $x = $ _____

2. $6x = 78$

    $x = $ _____

3. $8m = 256$

    $m = $ _____

4. $\frac{w}{3} = 8$

    $w = $ _____

5. $\frac{n}{8} = 12$

    $n = $ _____

6. $\frac{k}{5} = 75$

    $k = $ _____

7. $\frac{28}{t} = 4$

    $t = $ _____

8. $\frac{144}{x} = 9$

    $x = $ _____

9. $p\overline{)414}^{\,46}$

    $p = $ _____

Reteaching 5

Math Course 1, Lesson 5

- **Order of Operations, Part 1**

  - Work inside parentheses **first**.
  - Use that answer to finish working the problem from left to right.

    Example: 5 + (3 × 4)

    5 + 12 = 17

Practice:

Simplify 1–6.

1. 48 ÷ (12 ÷ 2) = _____

2. 40 − (20 − 12) = _____

3. 72 ÷ (3 × 3) = _____

4. 40 − 20 − 12 = _____

5. $\frac{(12)(24)}{8}$ = _____

6. 5 × 30 ÷ 15 = _____

Name _____

Reteaching 6

Math Course 1, Lesson 6

- **Fractional Parts**

  $\dfrac{\text{Numerator}}{\text{Denominator}}$ ← shows how many of the parts are counted
  ← shows the total number of parts

  Name **how many** parts are **shaded** out of a **total** number of parts.

   2 out of 3 are shaded $\left(\dfrac{2}{3}\right)$ $\dfrac{2}{3}$ (two thirds) is shaded
  $\dfrac{1}{3}$ (one third) is **not** shaded

  $\dfrac{1}{2}$ of a number ⟶ divide by 2

  $\dfrac{1}{3}$ of a number ⟶ divide by 3

  $\dfrac{1}{4}$ of a number ⟶ divide by 4

Practice:

1. What is the denominator of $\dfrac{13}{15}$? _____

2. What fraction of this circle is shaded? _____

3. What number is $\dfrac{1}{4}$ of 40? _____

4. What number is $\dfrac{1}{3}$ of 15? _____

5. What number is $\dfrac{1}{2}$ of $5.50? _____

6. A team won 3 of its 10 games.

   What fraction of games did the team win? _____

Name _____

Reteaching 7
Math Course 1, Lesson 7

- **Lines, Segments, and Rays**
- **Linear Measure**

Two systems of units are used to measure length:

- **U.S. Customary** (Uses fractions.)
  Some of the units in this system are inches, feet, yards, and miles.
- **Metric** (Uses decimals.)
  Some of the units in this system are millimeters, centimeters, meters, and kilometers.

Example: This line segment measures 25 mm on a metric ruler and about 1 in. on a customary ruler.

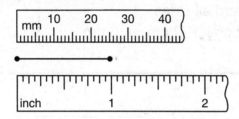

Practice:

1. Name each figure as a line, ray, or segment.

   a. ←——→   b. •—•   c. •——→

   _____   _____   _____

2. Which of the following illustrates a ray? _____

   A. A side of a rectangle

   B. The road between Austin and San Antonio

   C. A laser beam

3. How long is this line segment? _____

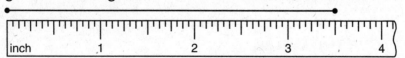

4. How long is this line segment? _____

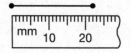

Name _____

Reteaching 8

Math Course 1, Lesson 8

- **Perimeter**

  - Perimeter is the distance **around** a shape.
  - Add all the sides.

    Examples:

    Perimeter ⟶ add all sides
    3 cm + 2 cm + 3 cm + 2 cm = 10 cm

    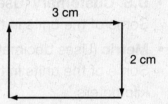

    The four sides of a square are equal in length.
    Perimeter ⟶ add all sides
    2 cm + 2 cm + 2 cm + 2 cm = 8 cm

    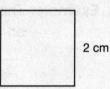

Practice:

1. What is the perimeter of this rectangle?

   _____

   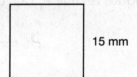

2. What is the perimeter of this square?

   _____

3. What is the perimeter of a square that has sides 5 inches long?

   _____

4. If the perimeter of a square is 16 inches, how long is each side?

   _____

Name _____

Reteaching 9
Math Course 1, Lesson 9

- # The Number Line: Ordering and Comparing

- Ordering: arranging numbers in order from least to greatest
- Comparing: showing that one number is equal to (=), less than (<), or greater than (>) another number

    Examples: $2 = 2 \quad 2 < 4 \quad 4 > 2$

    The smaller end (<) points to the *smaller* number (2 < 4).

Before comparing expressions, find the value of each expression.

$$\underbrace{2 + 4}_{6} \bigcirc \underbrace{2 \times 4}_{8}$$
$$6 < 8$$

Practice:

1. Arrange these amounts of money in order from least to greatest.
    $2.40 $24 $0.24

    _____

Compare 2–5.

2. $3 \times 5 \bigcirc 30 - 10$

3. $11 \times 7 \bigcirc 7 \times 11$

4. $8 - 4 \bigcirc 8 \div 4$

5. $9 - 2 \bigcirc 9 \times 2$

Name _____

Reteaching 10

Math Course 1, Lesson 10

- **Sequences**
- **Scales**

A **sequence** is an ordered list of numbers that follows a certain pattern. Find that pattern and continue it.
- **Addition sequence:** the same number is added to each term

    2, 4, 6, 8, 10, ... (+2 each term)

- **Multiplication sequence:** each term is multiplied by the same number

    2, 4, 8, 16, 32, ... (×2 each term)

- **Even** numbers end with 0, 2, 4, 6, or 8.
- **Odd** numbers end with 1, 3, 5, 7, or 9.

A **scale** is a display of numbers similar to a number line with an indicator to show the value of a certain measure. Find the value of the marks on the scale first. Then read the indicated number.

Practice:

1. What number is next in this sequence?

    3, 6, 12, 24, _____

2. What number is next in this sequence?

    7, 14, 21, 28, _____

3. What is the eighth number in this sequence?

    5, 10, 15, 20, ... _____

4. Which of the following numbers is odd? _____

    A. 329   **B.** 3246   **C.** 7890

5. What are the next three numbers in this sequence?

    4, 7, 10, 13, ____, ____, ____

6. What temperature is shown on this thermometer? _____

10   © Harcourt Achieve Inc. and Stephen Hake. All rights reserved.   Saxon Math Course 1

Name _____

Reteaching Inv. 1

Math Course 1, Investigation 1

- **Histograms**

A **histogram** is a special type of bar graph.
- Displays data in equal-size intervals
- Used to represent frequencies of events
- Each bar represents a range of values.
- No spaces between the bars
- Height of bar shows number of times data occur within the range.

Practice:

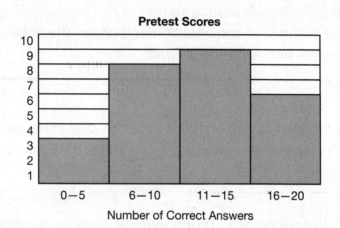

1. Which interval had the highest frequency of scores?

   _____

2. Which interval had the lowest frequency of scores?

   _____

3. What was the frequency of scores in the 16–20 range?

   _____

4. What was the frequency of scores above 10 correct answers?

   _____

Name _____

Reteaching 11

Math Course 1, Lesson 11

- **Problems About Combining**
- **Problems About Separating**

Combining stories have an **addition** pattern.
$$\text{Some} + \text{some more} = \text{total}$$

Separating stories have a **subtraction** pattern.
$$\text{Beginning amount} - \text{some went away} = \text{what remains}$$

Four-step process:
1. Read the problem and identify its pattern.
2. Write an equation for the given information.
3. Find the number that solves the equation and check the answer.
4. Review the question and write the answer.

Practice:

1. Shannon paused on step 125 on her way to the roof of the apartment building. If there are 180 steps in all, how many more steps did Shannon have to climb?

    _____

2. Luke has read 452 pages of a 705-page book. How many pages does Luke need to read to finish the book?

    _____

3. Kayla counted 73 red tomatoes in the garden. Her sister picked some of the tomatoes. When Kayla returned, she counted 25 red tomatoes. How many tomatoes did her sister pick?

    _____

4. Jordan's family is driving 230 miles to his grandmother's house. If they stop for lunch after 156 miles, how many more miles do they need to drive?

    _____

Name _____

Reteaching 12
Math Course 1, Lesson 12

- **Place Value Through Trillions**
- **Multistep Problems**

- To determine place value:

  A comma is always followed by 3 digits.

  Say the number in front of each comma followed by "trillion," "billion," "million," or "thousand."

Whole Number Place Value Chart

hundred trillions	ten trillions	trillions	,	hundred billions	ten billions	billions	,	hundred millions	ten millions	millions	,	hundred thousands	ten thousands	thousands	,	hundreds	tens	ones
Trillions				Billions				Millions				Thousands				Units (Ones)		

- To solve multistep problems:

  Find one number for each phrase with an operation.

  Then solve the simple problem.

Sum	The answer when we add
Difference	The answer when we subtract
Product	The answer when we multiply
Quotient	The answer when we divide

Example: When the **sum of 3 and 4** is subtracted from the **product of 3 and 4,** what is the difference?

$(3 \times 4) - (3 + 4) = 5$

Practice:

1. Write 5,000,000 in word form. _____

2. What is the place value of the 6 in 987,654,321? _____

3. What digit is in the millions place in 13,245,768? _____

4. Write the numeral for twenty-three million, four hundred two thousand.

   _____

5. What is the product of eight hundred forty and thirty-two?

   _____

6. What is the quotient when the sum of 9 and 6

   is divided by the difference of 9 and 6? _____

Saxon Math Course 1

Name _____

Reteaching 13
Math Course 1, Lesson 13

- **Problems About Comparing**
- **Elapsed-Time Problems**

- **Comparing** stories have a **subtraction** pattern.

  greater − lesser = difference

- **Elapsed-time** problems have a **subtraction** pattern.

  later − earlier = difference

Remember the **four-step process:**
1. Read the problem and identify its pattern.
2. Write an equation for the given information.
3. Find the number that solves the equation and check the answer.
4. Review the question and write the answer.

Practice:

1. The Mountain School has 865 students. The Lake School has 792 students. How many more students are in the Mountain School than in the Lake School?

   _____

2. How many years were there from 1776 to 1998?

   _____

3. How many years were there from 1865 to 1910?

   _____

4. Volume 1 has 1582 pages. Volume 2 has 1947 pages. How many more pages are in Volume 2?

   _____

Name _____

Reteaching 14
Math Course 1, Lesson 14

- **The Number Line: Negative Numbers**

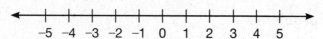

On the number line:

- **Positive** numbers are to the *right* of zero.
- **Negative** numbers are to the *left* of zero.
- **Zero** is neither positive nor negative.
- **Opposites** are numbers the same distance from zero (−5 and 5).
- **Integers** are all the counting numbers, their opposites, and zero.

Example: What number is 7 less than 3?

"7 less than 3" means to **start with 3** and subtract 7.

3 − 7

On the number line, start on 3 and count 7 integers to the left. The answer is −4.

Order matters in subtraction.

5 − 2 = 3 is different from 2 − 5 = −3

Reversing the order of subtraction results in the opposite answer.

Practice:

1. What number is 5 less than 3? _____

2. What number is 4 less than 0? _____

3. The morning temperature was −3°. The afternoon temperature was

   8 degrees higher. What was the afternoon temperature? _____

4. The morning temperature was −7°. By noon the temperature was 6°.

   How many degrees had the temperature risen? _____

Name _____

Reteaching 15

Math Course 1, Lesson 15

- **Problems About Equal Groups**

Equal groups word problems

Number of groups × number in group = total

Multiply to find the unknown total.

Divide to find an unknown factor.

Example: There were 232 students in 8 classrooms.

If there were the same number of students in each classroom, how many students would each classroom have?

- Write the equation.

  8 classrooms × n in each classroom = 232 students

- Divide to find the unknown factor.

  $$8\overline{)232} = 29 \text{ students}$$

Practice:

1. If 500 beads are put into bags of 20 beads, how many bags will be made?

   _____

2. The post office sold 39-cent stamps in sheets of 20 stamps. What was the price of a sheet of stamps?

   _____

3. If 750 pennies are put into rolls of 50 pennies each, how many rolls will there be?

   _____

4. David has 26 pages of trading cards in a notebook. Each page holds 9 trading cards. How many trading cards does David have in all?

   _____

Name _____

Reteaching 16
Math Course 1, Lesson 16

- **Rounding Whole Numbers**
- **Estimating**

- To round whole numbers:
    1. Underline the place value you are rounding to.
    2. Circle the digit to its right.
    3. Ask "Is the circled number 5 or more?"
       If so, add one to the underlined number.
       If not, the underlined number stays the same.
    4. Replace the circled number (and any numbers after it) with zero.

    Example: Round 472 to the nearest *hundred*.

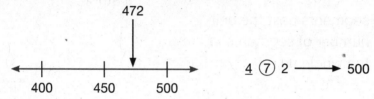

- To estimate answers, round the numbers before we add, subtract, multiply, or divide.

Practice:

1. Round 48,425 to the nearest thousand. _____

2. Round 5361 to the nearest hundred. _____

3. Estimate the product of 21 and 38. _____

4. Estimate the sum of 2345 and 6897 to the nearest thousand.

   _____

5. Estimate the difference of 642 and 357 to the nearest hundred.

   _____

6. The distance between New York, NY, and Los Angeles, CA, is 4685 km.

   Round that distance to the nearest thousand.

   _____

Saxon Math Course 1 © Harcourt Achieve Inc. and Stephen Hake. All rights reserved.

Name _____

Reteaching 17

Math Course 1, Lesson 17

- **Number Line: Fractions and Mixed Numbers**

- Point A represents the mixed number $2\frac{3}{5}$.

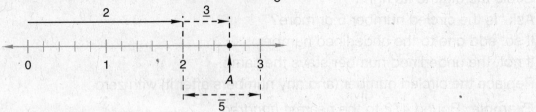

$2\frac{3}{5}$ ← segments past the unit
　　 ← number of segments in the whole unit
↑
number of units

- This inch ruler is divided into **sixteenths**. Count by sixteenths; reduce when possible.

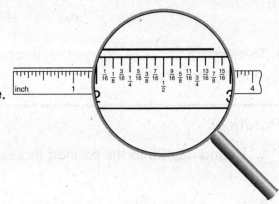

Practice:

1. Point B represents what mixed number on this number line? _____

Use a ruler to find the length of each of these line segments to the nearest sixteenth of an inch.

2. _____

3. _____

4. _____

Name _____

Reteaching 18
Math Course 1, Lesson 18

- **Average**
- **Line Graphs**

Average
- Add the numbers; then divide by the number of numbers.

   Example: What is the average of 8, 7, and 3?

   Add → 8 + 7 + 3 = 18   Divide → 18 ÷ 3 = 6

Halfway
- Add the two numbers; then divide by 2.

   Example: What number is halfway between 27 and 81?

   Add → 27 + 81 = 108   Divide by 2 → 108 ÷ 2 = 54

Line graphs
- Points connected by line segments; show how measurement changes over time

---

Practice:

1. What is the average of 3, 4, 6, 8, 8, and 7? _____

2. What number is halfway between 18 and 54? _____

3. What is the average of 167, 85, and 123? _____

Use the graph below to answer questions 4 and 5.

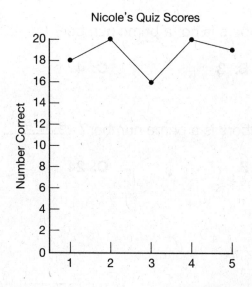

4. On which quiz did Nicole earn her lowest score? _____

5. What was Nicole's score on Quiz 4? _____

Saxon Math Course 1

Reteaching 19

Math Course 1, Lesson 19

- **Factors**
- **Prime Numbers**

- A **factor** of a number is a whole number that divides the number evenly.
  To list the factors of whole numbers:
  - *Start* with the number 1.
  - *End* with the number given.
  - Then find all the other *factors* of the number.
  - List the factors *in order*.

    Example: The factors of 12 are _1_ , _2_ , _3_ , _4_ , _6_ , _12_ .

- A **prime number** has exactly 2 different factors.
  The only factors are the number itself and 1.

    Example: 5 is only divisible by 5 and 1, so 5 is a prime number.
    6 is divisible by 1, 2, 3, and 6, so 6 is not a prime number.

Practice:

1. List the whole-number factors of 15. _____

2. List the six whole-number factors of 18. _____

3. List the whole-number factors of 21. _____

4. Which of these numbers is not a prime number? _____

   A. 2   B. 3   C. 4   D. 5

5. Which of these numbers is a prime number? _____

   A. 14   B. 17   C. 24   D. 25

Name _____

Reteaching 20

Math Course 1, Lesson 20

• **Greatest Common Factor (GCF)**

To find the greatest common factor (GCF):
- List (in order) the factors of the *smallest* number.
- Starting with the greatest factor, cross off factors that are not factors of the other numbers.
- Circle the greatest factor that is a factor of all the numbers. This is the **GCF**.

Example: Find the greatest common factor of 6, 9, and 15.

Factors of 6: 1, 2, 3, 6

6 is not a factor of 15 and 9: 1, 2, 3, 6̸

3 is a factor of 15 and 9: 1, 2, ③, 6̸

2 is not a factor of 15 and 9: 1, 2̸, ③, 6̸

1 is a factor of 15 and 9: 1, 2̸, ③, 6̸

Practice:

1. What is the greatest common factor (GCF) of 12 and 18? _____

2. What is the GCF of 15 and 21? _____

3. What is the GCF of 27 and 36? _____

4. What is the largest number that is a factor of both 24 and 36? _____

5. What is the GCF of 10, 15, and 20? _____

Name _____

Reteaching Inv.

Math Course 1, Investigation 2

- **Investigating Fractions with Manipulatives**

  - To subtract a fraction **from 1**, think of a shape divided into equal sections.

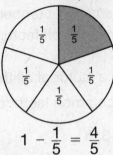

$$1 - \frac{1}{5} = \frac{4}{5}$$

Think about fractions as percents. A whole is 100%.

$\frac{1}{4} = 25\%$ $\frac{1}{2} = 50\%$ $\frac{3}{4} = 75\%$ $1 = 100\%$

- To **compare fractions**, compare pictures.

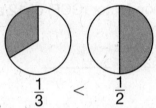

$\frac{1}{3} < \frac{1}{2}$

Practice:

Simplify 1–2.

1. $1 - \frac{2}{5} =$ _____

2. $1 - \frac{3}{8} =$ _____

3. What percent of a circle is $\frac{1}{2}$ of a circle? _____

4. Compare: $\frac{2}{3}$ ◯ $\frac{1}{2}$

5. Compare: $\frac{1}{4}$ ◯ $\frac{1}{2}$

6. Which percent best describes the shaded portion of this circle? _____

    A. 10%   B. 25%
    C. 50%   D. 80%

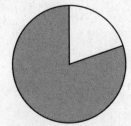

Name _____

Reteaching 21

Math Course 1, Lesson 21

- **Divisibility**

Last-Digit Tests
Inspect the last digit of the number. A number is divisible by . . .   2 if the last digit is even.   5 if the last digit is 0 or 5.   10 if the last digit is 0.

Sum-of-Digits Tests
Add the digits of the number and inspect the total. A number is divisible by . . .   3 if the sum of the digits is divisible by 3.   9 if the sum of the digits is divisible by 9.

Practice:

1. Which of these numbers is divisible by 2? _____

   A. 2612   B. 1541   C. 4263

2. Which of these numbers is divisible by 5? _____

   A. 1399   B. 1395   C. 1392

3. Which of these numbers is divisible by 3? _____

   A. 3456   B. 5678   C. 9124

4. Which of these numbers is divisible by 9? _____

   A. 6754   B. 8124   C. 7938

Saxon Math Course 1   © Harcourt Achieve Inc. and Stephen Hake. All rights reserved.   23

Name _____

Reteaching 22
Math Course 1, Lesson 22

- **"Equal Groups" Word Problems with Fractions**

What number is $\frac{3}{4}$ of 12?

Example:

1. Divide the total by the denominator (bottom number).

   $12 \div 4 = 3$

   $\frac{1}{4}$ of 12 is 3.

2. Multiply your answer by the numerator (top number).

   $3 \times 3 = 9$

   So, $\frac{3}{4}$ of 12 is 9.

Practice:

1. If $\frac{1}{3}$ of the 18 eggs were cracked, how many were not cracked?

   _____

2. What number is $\frac{2}{3}$ of 15? _____

3. What number is $\frac{3}{8}$ of 72? _____

4. How much is $\frac{5}{6}$ of two dozen? _____

5. Two fifths of the 40 answers were correct. How many answers were correct?

   _____

6. If $\frac{3}{4}$ of the 1000 show tickets were sold, how many tickets were sold?

   _____

Reteaching 23

Math Course 1, Lesson 23

- **Ratio**

A **ratio** is a way to describe a relationship between numbers. Each of the following forms is read the same: "Thirteen to fifteen."

$$13 \text{ to } 15 \qquad 13:15 \qquad \frac{13}{15}$$

- Write the ratio in the order asked.
- Reduce ratios if possible. $\frac{12}{16} = \frac{3}{4}$
- Leave ratios in fraction form. (Do not write a ratio as a mixed number.)

Practice:

1. What is the ratio of boys to girls in a class that has 15 boys and 16 girls?

   _____

2. A pet store has 16 dogs and 24 cats.

   What is the ratio of dogs to cats? _____

3. In a class of 25 students there are 12 boys.

   What is the ratio of boys to girls in the class? _____

4. The Bluebirds won 12 of their 20 games and lost the rest.

   What was the Bluebirds' win-loss ratio? _____

5. The Fireflies won 18 of their 24 games and lost the rest.

   What was the Fireflies' win-loss ratio? _____

Saxon Math Course 1

Name _____

Reteaching 24

Math Course 1, Lesson 24

- **Adding and Subtracting Fractions That Have Common Denominators**

Use fraction manipulatives to help you see that the denominator does not change.

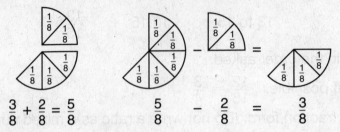

$$\frac{3}{8} + \frac{2}{8} = \frac{5}{8} \qquad \frac{5}{8} - \frac{2}{8} = \frac{3}{8}$$

Notice that the denominators are the same. So, add or subtract the numerators. The denominator does **not** change.

Practice:

Simplify 1–6.

1. $\frac{3}{5} + \frac{1}{5} =$ _____

2. $\frac{2}{3} + \frac{1}{3} =$ _____

3. $\frac{1}{6} + \frac{2}{6} + \frac{2}{6} =$ _____

4. $\frac{7}{8} - \frac{2}{8} =$ _____

5. $\frac{6}{7} - \frac{6}{7} =$ _____

6. $\frac{4}{9} + \frac{5}{9} - \frac{2}{9}$ _____

Name _____

Reteaching 25
Math Course 1, Lesson 25

- **Writing Division Answers as Mixed Numbers**
- **Multiples**

- We write some division answers as mixed numbers.
- A mixed number is a whole number plus a fraction.

   whole number → $2\frac{3}{4}$ ← fraction

   mixed number

- Put the remainder over the divisor to make the fraction of a mixed number.

   Example 1: Write $\frac{25}{6}$ as a mixed number.

   $$\begin{array}{r} 4\frac{1}{6} \\ 6\overline{)25} \\ \underline{24} \\ 1 \end{array}$$

   Example 2: A whole circle is 100% of a circle.
   One third of a circle is what percent of a circle?

   $$\begin{array}{r} 33\frac{1}{3}\% \\ 3\overline{)100\%} \\ \underline{9} \\ 10 \\ \underline{9} \\ 1 \end{array}$$

- We find **multiples** of a number by **multiplying** the number by 1, 2, 3, 4, 5, 6, and so on.

   Example: The first six multiples of 3 are 3, 6, 9, 12, 15, and 18.

Practice:

1. Write $\frac{11}{3}$ as a mixed number. _____

2. A 25-inch length of string was cut into 4 equal lengths.

   How long was each piece of string? _____

3. Add and convert to a mixed number: $\frac{2}{5} + \frac{4}{5} =$ _____

4. What are the first six multiples of 4? _____

5. What are the first five multiples of 10? _____

Name _____

Reteaching 26
Math Course 1, Lesson 26

- **Using Manipulatives to Reduce Fractions**
- **Adding and Subtracting Mixed Numbers**

- Each picture illustrates half of a circle that has been divided a different way.

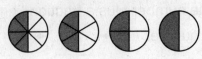

The picture with the fewest pieces is $\frac{1}{2}$. Each of the other fractions $\left(\frac{4}{8}, \frac{3}{6}, \frac{2}{4}\right)$ **reduces** to $\frac{1}{2}$.

We **reduce** when we use the *fewest* number of fraction pieces to show the amount.

We **convert** when we change a fraction to a mixed number or a whole number.

- When adding mixed numbers:
  1. Add the fractions.
  2. Add the whole numbers.
  3. Reduce or convert the fraction if needed.

Practice:

Reduce 1–3.

1. $\frac{6}{8}$ _____

2. $\frac{6}{9}$ _____

3. $\frac{6}{10}$ _____

4. Add and convert: $2\frac{2}{6} + 3\frac{5}{6} =$ _____

5. $5\frac{3}{4} - 2\frac{1}{4} =$ _____

6. $\quad 2\frac{1}{3}$
   $+\ 4\frac{2}{3}$
   ―――

7. $3 + N = 6\frac{1}{4}$

   $N =$ _____

Name _____

Reteaching 27

Math Course 1, Lesson 27

• **Measures of a Circle**

• The **circumference** is the *distance around* the circle (perimeter).
• The **diameter** is the *distance across a circle* through the center.
• The **radius** is the *distance from the center to the circle*.
• The diameter is twice the radius.

$$d = 2r$$

• The radius is half the diameter.

$$r = \frac{1}{2}d$$

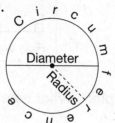

Practice:

1. Which of these words names the distance from the center to the circle? _____

   A. radius   **B.** circumference   **C.** diameter

2. The diameter of a bicycle tire is 22 inches.

   What is the radius of the tire? _____

3. Which segment in the circle at right is a radius? _____

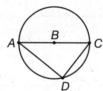

   A. segment AB   **B.** segment AC   **C.** segment AD   **D.** segment CD

4. The diameter of a big circle is 30 feet.

   What is the ratio of the circle's radius to its diameter? _____

5. If the radius of a circle is 18 cm, what is its diameter? _____

Name _____

Reteaching 28

Math Course 1, Lesson 28

- **Angles**

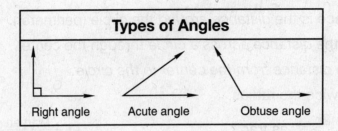

Example: What type of angle is ∠y?

What type of angle is ∠x?

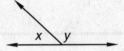

Two ways of naming angles:
- Use the letter of the vertex: ∠Q
- Use 3 letters with the vertex in the middle:
  ∠PQR or ∠RQP

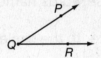

Practice:

1. Name the obtuse angle in triangle ABC.

   _____

2. What kind of angle is ∠A? _____

3. Write three names for this angle?

   _____

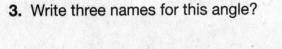 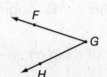

4. What kind of angle is formed by the hands of a clock showing 3 o'clock?

Name _____

Reteaching 29
Math Course 1, Lesson 29

- **Multiplying Fractions**
- **Reducing Fractions by Dividing by Common Factors**

Multiplying fractions
- "Of" means "multiply."

    Multiply numerator × numerator and multiply denominator × denominator.

- Set up whole numbers as fractions. $\left(4 = \frac{4}{1}\right)$

    Example: $\frac{2}{3}$ of $\frac{4}{5}$ means $\frac{2}{3} \times \frac{4}{5} = \frac{8}{15}$

    Example: $4 \times \frac{2}{3}$

    $\frac{4}{1} \times \frac{2}{3} = \frac{8}{3} = 2\frac{2}{3}$

Reducing fractions
Find a number that will **evenly** divide **both** the numerator and the denominator. (the GCF)

Example: $\frac{4}{6} \div 2 = \frac{2}{3}$

Practice:

Simplify 1–2.

1. $\frac{1}{2}$ of $\frac{3}{5}$ = _____

2. $2 \times \frac{2}{3}$ = _____

3. Convert $\frac{8}{6}$ to a mixed number and reduce the fraction. _____

4. What is the product of $\frac{1}{2}$ and $\frac{2}{3}$? _____

5. Amanda correctly answered 18 of the 20 questions.

    What fraction of the questions did she answer correctly? _____

6. What is the ratio of chickens to ducks

    with 24 chickens and 18 ducks? _____

Saxon Math Course 1

Name _____

Reteaching 30

Math Course 1, Lesson 30

- **Least Common Multiple (LCM)**
- **Reciprocals**

Multiples remind us of multiplication.
- Multiples: Think times table.

    Multiples of 3: 3, 6, 9, (12), 15, 18, 21, (24), . . .
    Multiples of 4: 4, 8, (12), 16, 20, (24), 28, . . .
    12 and 24 are **common** multiples.
    12 is the **least** common multiple (LCM).

Reciprocal $\longrightarrow$ "flip" the fraction

   Example: $4 = \frac{4}{1}$, so the reciprocal of 4 is $\frac{1}{4}$.

- **Rule:** The product of any fraction and its reciprocal equals 1.

    Examples: $\frac{3}{4} \times \frac{4}{3} = \frac{12}{12} = 1$ $\frac{7}{9} \times \frac{9}{7} = \frac{63}{63} = 1$

Practice:

1. What is the least common multiple (LCM) of 4 and 5? _____

2. What is the LCM of 2 and 7? _____

3. What is the least common multiple (LCM) of 3, 6, and 9? _____

4. What is the reciprocal of $\frac{3}{7}$? _____

5. $\frac{3}{5} \times \boxed{} = 1$

6. How many $\frac{4}{5}$'s are in 1? _____

Name _____

Reteaching Inv. 3

Math Course 1, Investigation 3

• Measuring and Drawing Angles with a Protractor

Angles can be measured in degrees. Use a protractor to measure angles.

∠EOA is an acute angle.
An acute angle is less than 90°.

∠FOA is a right angle.
A right angle is 90°.

∠HOA is an obtuse angle.
An obtuse angle is more than 90° and less than 180°.

∠IOA is a straight angle.
A straight angle is 180°.

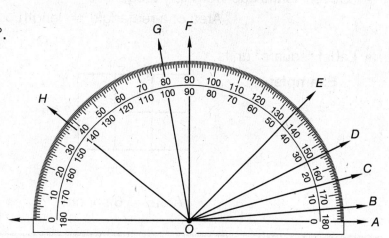

Practice:

Use a protractor to help answer each question.

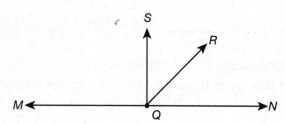

1. Which angle above could measure 45°? _____

2. What is the measure of ∠MQR? _____

3. Which angle measures 180°? _____

4. Which angle is a 90° angle? _____

Saxon Math Course 1

Name _____

Reteaching 3
Math Course 1, Lesson 31

- **Areas of Rectangles**

When we measure the "inside" of a flat shape, we measure **area.**
- "Cover" is the cue word for "area."

Area of a rectangle = length × width

- Label "square" units.

Example:

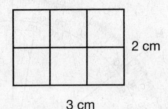

3 cm

Area = 6 sq. cm

Practice:

1. How many square tiles with sides 1 ft long would be needed to cover the floor of an 8-ft-by-10-ft room?

   _____

2. A rectangle is 5 inches long and 7 inches wide. How many one-square-inch tiles are needed to cover its area?

   _____

3. What is the area of this rectangle?

   16 ft
   14 ft

   _____

4. How many square centimeters are needed to cover this rectangle?

   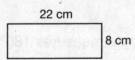

   _____

5. The perimeter of a square is 12 inches. What is its area?

   _____

34 © Harcourt Achieve Inc. and Stephen Hake. All rights reserved. Saxon Math Course 1

Name _____

Reteaching 32
Math Course 1, Lesson 32

- **Expanded Notation**
- **More on Elapsed Time**

- To write in expanded notation:
  Name the **place value** of each digit:
  Example: $3256 = (3 \times 1000) + (2 \times 100) + (5 \times 10) + (6 \times 1)$

- When given expanded notation:
  1. Count the places in the first parentheses:
     $(4 \times 1000) + (6 \times 10) + (2 \times 1)$
  2. Draw digit lines for each place: ___, ___ ___ ___
  3. Fill in the digit lines: _4_ _0_ _6_ _2_

• Do elapsed time problems in three steps.
 	
  1. State the "time now."
  2. Count the hours forward or backward.
  3. Count the minutes forward or backward.
	• Measure time between two stated times in two steps.
	
	  1. Count the hours forward or backward.
	  2. Count the minutes forward or backward.

Example: What time is $2\frac{1}{2}$ hours after 10:43 a.m.? Time now → 10:43 a.m. Count forward 2 hr → 12:43 p.m. Count forward 30 min → 1:13 a.m.	**Example:** How long is it from 7:15 a.m. to 10:10 a.m.? 7:15 a.m. to 9:15 a.m. → 2 hours 9:15 a.m. to 10:10 a.m. → 55 minutes It is 2 hours and 55 minutes from 7:15 a.m. to 10:10 a.m.

Practice:

1. Write $(4 \times 100) + (6 \times 1)$ in standard notation. _____

2. Write $(7 \times 1000) + (3 \times 100)$ in standard notation. _____

3. How long is it from 6:45 a.m. to 10:15 a.m.?

   _____

4. The movie starts at 1:10 p.m. and ends at 3:22 p.m. How long is the movie?

   _____

Name _____

Reteaching 33

Math Course 1, Lesson 33

• **Writing Percents as Fractions, Part 1**

Percents may be written as fractions.
  • The "percent" is the numerator; the denominator is 100.
  • Cancel the matching zeros (when possible).
  • Reduce as necessary.
    Example: $40\% = \frac{40}{100} = \frac{2}{5}$

Practice:

1. Write 40% as a reduced fraction. _____

2. Write 60% as a reduced fraction. _____

3. Write 75% as a reduced fraction. _____

4. Write 12% as a reduced fraction. _____

5. Twenty percent of the answers were wrong.

   What fraction of the answers were wrong? _____

6. Fifty percent of the answers were wrong.

   What fraction of the answers were wrong? _____

36   © Harcourt Achieve Inc. and Stephen Hake. All rights reserved.   Saxon Math Course 1

Name _____

Reteaching 34
Math Course 1, Lesson 34

• Decimal Place Value

We can use bills and coins to help us understand place value.
- As we move to the right, each place is one tenth of the value of the place to its left.

hundreds place	tens place	ones place	.	tenths place	hundredths place	thousandths place
$100 bills	$10 bills	$1 bills	.	dimes	pennies	mills

- The "-ths" ending indicates the place value is to the right of the decimal point.

Practice:

1. Which digit in 6.543 is in the thousandths place? _____

2. Which digit is in the tenths place in 5467.982? _____

3. Write the decimal numeral for two and fourteen hundredths. _____

4. Which digit in 34.76 has the same place value as the 1 in 8.219? _____

5. Which digit in 135.29 has the same place value as the 4 in 37.48? _____

6. Which digit in 2.819 has the same place value as the 7 in 6.537? _____

Reteaching 35

Math Course 1, Lesson 35

- **Writing Decimal Numbers as Fractions, Part 1**
- **Reading and Writing Decimal Numbers**

- Notice that the number of decimal places in the decimal number equals the number of zeros in the denominator.

  $0.3 = \frac{3}{10}$ $0.21 = \frac{21}{100}$ $0.023 = \frac{23}{1000}$

- To read a mixed decimal number, we read the whole number part, say "and," and then read the decimal part.

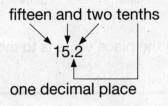

Practice:

1. Write 0.28 as a common fraction. _____

2. Write $\frac{7}{100}$ in decimal form. _____

3. Write $\frac{56}{1000}$ in decimal form. _____

4. Write the decimal numeral for nineteen hundredths. _____

5. Write the decimal numeral for seven and three tenths. _____

6. Write the words for 4.21. _____

Reteaching 36

Math Course 1, Lesson 36

- **Subtracting Fractions and Mixed Numbers from Whole Numbers**

- Rename 1 from the whole number as a fraction with the same denominator number to change the whole number to a mixed number.
- Subtract the fractions. Then subtract the whole numbers.

Example: $4 - \frac{1}{2}$ ⟶ Rename 4 as $3\frac{2}{2}$ ⟶ $\begin{array}{r} 3\frac{2}{2} \\ - \frac{1}{2} \\ \hline 3\frac{1}{2} \end{array}$

Practice:

Simplify 1–6.

1. $2 - \frac{1}{4}$ = _____

2. $4 - \frac{3}{8}$ = _____

3. $5 - \frac{3}{10}$ = _____

4. $3 - 1\frac{2}{3}$ = _____

5. $6 - 2\frac{3}{5}$ = _____

6. $10 - 5\frac{5}{6}$ = _____

Name _____

Reteaching 37

Math Course 1, Lesson 37

• **Adding and Subtracting Decimal Numbers**

We set up decimal numbers for addition or subtraction by lining up the decimal points. Empty places are treated as zeros.

```
    2.3
    2.41
+  31.7
```

When we line up the decimal points, we also align digits that have the same place values.

Remember to subtract in order.

Practice:

Simplify 1–6.

1. 0.6 + 0.9 = _____

2. 0.24 − 0.12 = _____

3. 1.5 + 2.34 = _____

4. 3.86 − 0.7 = _____

5. 0.87 + 6.1 = _____

6. 2.00 − 0.90 = _____

Name _____

Reteaching 38
Math Course 1, Lesson 38

- **Adding and Subtracting Decimal Numbers and Whole Numbers**
- **Squares and Square Roots**

- To line up decimal points when adding or subtracting decimal numbers:
    1. Place the decimal point **after** the whole number.
    2. Fill empty places with zeros. Borrow as necessary.

    Example: $3 - 1.2$ $\overset{21}{\cancel{3}.0}$
    $\underline{- 1.2}$
    1.8

- We square a number by multiplying the number by itself.

    5^2 means "five squared" → $5 \times 5 = 25$
    Five squared equals 25.

- Finding the square root is the inverse of squaring a number.

    $\sqrt{100}$ means "the square root of 100"
    Think: What number multiplied by itself equals 100?
    $10 \times 10 = 100$
    The square root of 100 is **10**.

Practice:

Simplify 1–6.

1. $9.12 - 7.8 =$ _____

2. $2.2 + 0.63 + 7 =$ _____

3. $14 + 9.5 + 16.8 =$ _____

4. $0.62 + (3 - 0.14) =$ _____

5. $3^2 + \sqrt{25} =$ _____

6. $3\frac{2}{3} + \sqrt{16} =$ _____

Name _____

Reteaching 3

Math Course 1, Lesson 39

- **Multiplying Decimal Numbers**

To multiply decimal numbers:
1. Multiply.
2. Count the digits to the **RIGHT** of the decimal points in all the factors.
3. Place the decimal point in the product that many places from the right-hand side.

Example:
$$\begin{array}{r} 0.15 \\ \times\ 0.9 \\ \hline 0.135 \end{array} \Big\} 3 \text{ places}$$

Practice:

Simplify 1–4.

1. 0.5 × 0.14 = _____

2. 1.6 × 0.9 = _____

3. 0.24 × 0.13 = _____

4. 0.3 × 0.8 × 0.2 = _____

5. What is the product of 2.3 and 0.024? _____

Name _____

Reteaching 40

Math Course 1, Lesson 40

- **Using Zero as a Place Holder**

When adding and subtracting decimal numbers, we line up the decimals points.
- Fill each empty decimal place with a zero.

Examples:

$$0.5 - 0.32 \longrightarrow \begin{array}{r} 0.5\_ \\ -\ 0.32 \\ \hline \end{array} \longrightarrow \begin{array}{r} 0.50 \\ -\ 0.32 \\ \hline \end{array}$$

$$3 + 0.4 \longrightarrow \begin{array}{r} 3.\_ \\ +\ 0.4 \\ \hline \end{array} \longrightarrow \begin{array}{r} 3.0 \\ +\ 0.4 \\ \hline \end{array}$$

Practice:

Simplify 1–6.

1. 2 − 0.53 = _____

2. 1.7 − 0.49 = _____

3. 8.6 − 2.92 = _____

4. 6 − (0.5 + 0.8) = _____

5. 7 − (3.4 + 0.91) = _____

6. 4 + (2.1 − 0.73) = _____

Saxon Math Course 1

Name _____

Reteaching Inv.

Math Course 1, Investigation 4

- **Data Collection and Surveys**

- **Quantitative** data: numbers
- **Qualitative** data: categories
- **Closed-option** surveys: limit responses–multiple choice
- **Open-option** surveys: do not limit choices
- Surveys gather data about a **population**, or certain group of people.
- A **representative sample** should have characteristics similar to the entire population.

Practice:

1. Meghan asked each of her classmates what their favorite animal was. Was she collecting quantitative or qualitative data?

   _____

2. Jasmine counted the number of raisins in five different bowls of cereal. Was she collecting quantitative or qualitative data?

   _____

3. Tim showed a list of colors when surveying his friends about what color to paint his room. What kind of survey was it?

   _____

4. A company wants to know what kinds of clothes students in the middle school like. Should their representative sample include first graders or seventh graders?

   _____

Name _____

Reteaching 41

Math Course 1, Lesson 41

- **Finding a Percent of a Number**

  - To find a percent of a number:
    1. Change the percent to a fraction or decimal.
    2. Then multiply.

    Example: What number is 75% of 20?

    Change the percent to a fraction; then multiply.

    $\frac{75}{100}$ reduces to $\frac{3}{4}$ $\frac{3}{4}$ of 20

    $$\frac{3}{\overset{1}{\cancel{4}}} \times \overset{5}{\cancel{20}} = 15$$

    Change the percent to a decimal; then multiply.

    $$\begin{array}{r} 0.75 \\ \times\ 20 \\ \hline 15.00 \end{array}$$

  - To find a total price for a purchase:
    1. Find the sales tax on the purchase.
    2. Add the tax amount to the purchase price.

Practice:

1. Write 60% as a reduced fraction and as a decimal numeral. _____

2. Matthew answered 40% of the 50 questions correctly. How many questions did he answer correctly?

   _____

3. A $40 pair of shoes is on sale for 25% off the regular price.

   How much money is 25% of $40? _____

4. The sales-tax rate was 8%. Brianna bought a craft kit for $9.49.

   How much was the tax on the kit? _____

5. What is the total price of a $22.95 item plus 6% sales tax? _____

6. Sam ordered a $4.90 meal. The tax rate was 7%. He paid with a $10 bill.

   How much money should he get back? _____

Saxon Math Course 1

Name _____

Reteaching 42

Math Course 1, Lesson 42

- **Renaming Fractions by Multiplying by 1**

  - To **rename** a fraction means to make an **equivalent fraction.**
    Multiply the fraction by a fraction equal to 1. The new fraction will be an equivalent fraction.

    Example: Write a fraction equal to $\frac{1}{2}$ that has a denominator of 20.

    $$\frac{1}{2} = \frac{?}{20} \qquad \frac{1}{2} \times \frac{10}{10} = \frac{10}{20}$$

  - Try the shortcut: divide; then multiply.

    Example: $\times \frac{2}{3} = \frac{?}{15}$ $\qquad (15 \div 3) \times 2 = 10$

Practice:

Solve 1–4.

1. $\frac{3}{4} = \frac{?}{16} =$ _____

2. $\frac{2}{7} = \frac{?}{21} =$ _____

3. $\frac{8}{25} = \frac{?}{100} =$ _____

4. $\frac{7}{20} = \frac{?}{100} =$ _____

5. By what fraction equal to 1 should $\frac{3}{8}$ be multiplied to form $\frac{6}{16}$? _____

6. Write $\frac{1}{4}$ as a fraction with 8 as the denominator.

   Then add the fraction to $\frac{1}{8}$. What is the sum? _____

Name _____

Reteaching 43

Math Course 1, Lesson 43

• Unknown-Number Problems with Fractions and Decimals

Find unknown numbers in fraction and decimal problems just like finding unknown numbers in whole-number problems.

Addition
- To find an unknown **addend** ⟶ subtract

  Example: $\frac{1}{5} + a = \frac{4}{5}$ ⟶ $\frac{4}{5} - \frac{1}{5} = a$ ⟶ $a = \frac{3}{5}$

Subtraction
- To find the unknown **minuend** (first number in subtraction) ⟶ add

  Example: $n - 3 = 2.2$ ⟶ $2.2 + 3 = n$ ⟶ $n = 5.2$

- To find the unknown **subtrahend** ⟶ subtract

  Example: $5.3 - y = 2$ ⟶ $5.3 - 2 = y$ ⟶ $y = 3.3$

Multiplication
- For a product of 1, the two factors must be reciprocals.

  Example: $\frac{3}{5}f = 1$ ⟶ $\frac{3}{5} \times \frac{5}{3} = \frac{15}{15} = 1$ ⟶ $f = \frac{5}{3}$

  Check your work by replacing the letter with the answer.

Practice:

Solve 1–6.

1. $4.56 + n = 7$

   n = _____

2. $1 - x = 0.58$

   x = _____

3. $w + \frac{5}{7} = 1\frac{3}{7}$

   w = _____

4. $5\frac{5}{8} - m = 2$

   m = _____

5. $11.4 - y = 6.2$

   y = _____

6. $\frac{4}{7}z = 1$

   z = _____

Name _____

Reteaching 4
Math Course 1, Lesson 44

- **Simplifying Decimal Numbers**
- **Comparing Decimal Numbers**

- To simplify decimal numbers:
    1. Remove the extra zeros in **front** and **back.**
    2. Keep the zero in front of the decimal.
    3. Do not remove a zero between other digits.
        Examples:
        $\cancel{0}2.01\cancel{0}\cancel{0} = 2.01$
        $0.42\cancel{0}\cancel{0} = 0.42$

- To compare decimal numbers:
    1. Ignore zeros at the end of a decimal number.
       They do not change the value of the number.
       $0.3 = 0.300 = 0.3000$
    2. Write the numbers with the same number of decimal places.
       $0.3 \bigcirc 0.209 \longrightarrow 0.300 \bigcirc 0.209$
    3. Compare the greatest place value first.
       $0.\underline{3}00 > 0.\underline{2}09$

Practice:

Simplify 1–2.

1. 3.0400 = _____

2. 0. 09100 = _____

Compare 3–5.

3. 2.45 ◯ 0.425

4. 0.081 ◯ 0.81

5. 0.5 + 0.5 ◯ 0.5 × 0.5

Reteaching 45

Math Course 1, Lesson 45

• **Dividing a Decimal Number by a Whole Number**

To divide decimal numbers by a whole number:
1. Put the **first** number **inside** the division box.
2. Put the **second** number **in front** of the box.
3. Put the decimal straight up on the answer line.
4. Use zero as a placeholder.
5. Put a digit above each digit.
6. Add zeros to the dividend and keep dividing until there is no remainder (or until digits repeat).

Example: 0.3 ÷ 4

4)0.3

$0.$
4)0.3

0.075
4)0.300

> When dividing by a **whole** number, decimal goes straight **up**.

Practice:

Simplify 1–6.

1. 5)1.8

2. 6)0.27

3. 9)2.43

4. 0.62 ÷ 2 = _____

5. 0.432 ÷ 8 = _____

6. 3.9 ÷ 4 = _____

Name _____

Reteaching 46

Math Course 1, Lesson 46

• Mentally Multiplying Decimal Numbers by 10 and by 100

When we multiply decimal numbers by 10 or by 100, the digits shift to the left. When the digits shift left, the decimal point is shifting to the right.
- To multiply by **10**: Shift the decimal to the **right** *one* place.
- To multiply by **100**: Shift the decimal to the **right** *two* places.

Examples: Shift right ⟶

$$1.234 \times 100 = 123.4$$

$$7.8 \times 100 = 780$$

Remember: If multiplying by a whole number, the product will be larger than the starting number.

Practice:

1. Mentally calculate this product:

    6.5 × 100 = _____

Simplify 2–6.

2. 87.56 × 10 = _____

3. 35.79 × 10 = _____

4. 2.4 × 100 = _____

5. 0.81 × 10 = _____

6. 0.6 × 100 = _____

Name _____

Reteaching 47
Math Course 1, Lesson 47

- **Circumference**
- **Pi (π)**

If the **radius** or **diameter** of a circle is known, the **circumference** can be found.

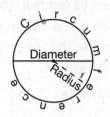

Multiply the diameter by π: $C = \pi d$
Use 3.14 as an approximation for π: $C = 3.14d$

Remember: $d = 2r$
So, $C = 2\pi r$

Practice:

Solve. Use 3.14 for π.

1. The diameter of a tortilla is 10 inches. What is the circumference of the tortilla?

   _____

2. The diameter of a circle is 200 mm. What is the circumference of the circle?

   _____

3. Marco used a compass to draw a circle with a radius of 5 cm.

   What was the circumference of the circle? _____

4. The diameter of a truck's tire is 40 inches. How far down the road will the tire travel if it makes one full turn? (Round your answer to the nearest inch.)

   _____

5. What is the circumference of the circle?

   _____

Name _____

Reteaching 48

Math Course 1, Lesson 48

• Subtracting Mixed Numbers with Regrouping, Part 1

If you need to regroup (borrow):
1. Rename one of the wholes.
2. Combine the fraction with the renamed 1.
3. Then subtract.

Example: $5\frac{1}{3}$ $\quad 4 + \frac{3}{3} + \frac{1}{3} \longrightarrow \quad 4\frac{4}{3}$
$\quad\quad\quad\quad -2\frac{2}{3}$ $\quad\quad\quad\quad\quad\quad\quad\quad\quad -2\frac{2}{3}$
$\quad\quad\quad\quad\quad\quad\quad\quad\quad\quad\quad\quad\quad\quad\quad\quad\quad\quad\quad 2\frac{2}{3}$

Practice:

Simplify 1–5.

1. $5\frac{1}{7}$
    $-2\frac{3}{7}$

2. $6\frac{1}{3}$
    $-2\frac{2}{3}$

3. $4\frac{2}{4}$
    $-1\frac{3}{4}$

4. $3\frac{2}{5} - 1\frac{3}{5} =$ _____

5. $7\frac{3}{8} - 2\frac{5}{8} =$ _____

Name _____

Reteaching 49

Math Course 1, Lesson 49

- **Dividing by a Decimal Number**

Change the problem to division by a whole number:
1. Move the decimal in the divisor **over** to make a whole number.
2. Move the decimal in the dividend **over** the same number of places.
3. Put a decimal in the quotient **up** above the moved decimal.
4. Use zero as a placeholder.
5. Put a digit above each digit.
6. Add zeros to the dividend and keep dividing until there is no remainder (or until digits repeat).

Example: $0.4 \overline{) 0.1300}$ → quotient $0 0.325$ (up, over over)

Practice:

Simplify 1–5.

1. $0.5 \overline{)0.018}$

2. $0.4 \overline{)42}$

3. $0.7 \overline{)0.91}$

4. $0.459 \div 0.09 = $ _____

5. $6 \div (0.8 \times 3) = $ _____

Saxon Math Course 1 © Harcourt Achieve Inc. and Stephen Hake. All rights reserved.

Name _____

Reteaching 50
Math Course 1, Lesson 50

- **Decimal Number Line (Tenths)**
  - We can locate decimal numbers on the number line.
  - On the number line, the distance between consecutive whole numbers is divided into ten equal lengths.
  - Each length is $\frac{1}{10}$.
  - Point y is 4 marks beyond the 7. So y is on $7\frac{4}{10}$.
  - We can rename $7\frac{4}{10}$ as 7.4.

    Example: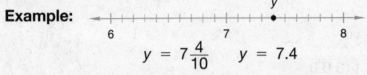

    $y = 7\frac{4}{10}$ $y = 7.4$

Practice:

1. To which decimal number is the arrow pointing? _____

2. To which decimal number is the arrow pointing? _____

   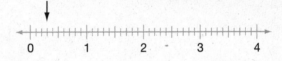

3. To which decimal number is the arrow pointing? _____

4. To which decimal number is the arrow pointing? _____

Name _____

Reteaching Inv. 5

Math Course 1, Investigation 5

- **Displaying Data**

Qualitative data can be displayed in
- a bar graph
- a pictograph
- a circle graph (or pie chart)

Quantitative data can be displayed in
- a histogram
- a line plot
- stem-and-leaf plots

Describe data in terms of mean, median, mode, and range.

Practice:

1. Mark's first seven test scores are shown below.

   What is the median of these scores? _____

   95, 100, 80, 95, 85, 80, 80

2. What is the mean of Mark's test scores shown above?

   (Round to the nearest whole number.) _____

3. Darcy surveyed the students in her class to find out how many had brothers and sisters. She found that $\frac{1}{2}$ had only 1 brother or sister, $\frac{1}{4}$ had more than 1 brother or sister, and $\frac{1}{4}$ had no brothers or sisters. Display Darcy's data in a circle graph.

4. Based on Darcy's survey, do more students have only one brother or sister or more than one brother or sister? _____

Name _____

Reteaching 5

Math Course 1, Lesson 51

- **Rounding Decimal Numbers**

To round decimal numbers:
- Circle the place value you are rounding to.
- Underline the digit to its right.
- If the underlined number is 5 or more, add 1 to the circled number.
  If the underlined number is 4 or less, the circled number stays the same.
- Drop all digits after the circled digit.

Example: Round to the hundredths' place: 1.6⑦85 ⟶ 1.68

Practice:

1. Round 2.357 to the nearest tenth. _____

2. Round 0.546 to the nearest tenth. _____

3. Round 3.6875 to the nearest hundredth. _____

4. Round 0.942 to the nearest hundredth. _____

5. Divide $8.91 by 6 and round the quotient to the nearest cent. _____

6. Divide $4.70 by 7 and round the quotient to the nearest cent. _____

Name _____

Reteaching 52

Math Course 1, Lesson 52

• Mentally Dividing Decimal Numbers by 10 and by 100

Since we are dividing, the answer will be less than the starting number.
- Shift the decimal **left** for division—one place for each **zero** in the divisor.
  Fill empty places with zeros.

Example: ← Shift left

$$3.4 \div 10 = 0.34$$
$$3.4 \div 100 = 0.034$$

Practice:

Simplify 1–6.

1. $95.4 \div 10 =$ _____

2. $3.8 \div 10 =$ _____

3. $71.5 \div 100 =$ _____

4. $3.6 \div 100 =$ _____

5. $2.25 \div 100 =$ _____

6. $87.9 \div 100 =$ _____

Name _____

Reteaching 53

Math Course 1, Lesson 53

- **Decimals Chart**
- **Simplifying Fractions**

Decimal Reminders

Operation	+ or −	×	÷ by whole (W)	÷ by decimal (D)
Memory cue	line up $\underline{\begin{array}{r}.\\+.\end{array}}$ $.$	×; then count $.\,\_\,\_$ $\underline{\times.\,\_}$ $\_\,\_\,\_$	up $W\overline{)\,\overset{.}{}\!\!.}$	over, over, up $D.\overline{)\,\underset{\smile}{}}$

You may need to:
- Place a decimal point to the right of a whole number.
- Fill empty places with zeros.

To reduce fractions to lowest terms:
- Divide both numbers by the largest number that will go into both evenly. (This is the GCF.)

Example: $\frac{4}{8}$ Divide the numerator and denominator by 4 (not 2).

$$\frac{4 \div 4}{8 \div 4} = \frac{1}{2}$$

To convert and reduce improper fractions to mixed numbers:
- Fix the "top heavy" number.
- Add it to the whole number.

Example: $2\frac{10}{8}$ Fix the "top heavy" number.

$$\frac{10}{8} = 1\frac{2}{8} = 1\frac{1}{4}$$

Add to the whole number.

$$2 + 1\frac{1}{4} = 3\frac{1}{4}$$

If you need to reduce **and** convert you may do so in either order:

Practice:

1. What do you do with empty places when dividing with decimals?

   _____

Simplify 2–5.

2. $\frac{3}{8} + \frac{7}{8} =$ _____

3. $\frac{5}{12} + \frac{11}{12} =$ _____

4. $\frac{3}{10} + \frac{9}{10} =$ _____

5. $\frac{7}{9} + \frac{5}{9} =$ _____

Name _____

Reteaching 54
Math Course 1, Lesson 54

- **Reducing by Grouping Factors Equal to 1**

Notice that some factors appear in both the dividend and the divisor. Since $2 \div 2 = 1$, we can mark combinations of factors equal to 1. Then solve the simplified problem.

Example: $\dfrac{\cancel{2} \cdot \cancel{2} \cdot \cancel{3} \cdot 5}{\cancel{2} \cdot \cancel{2} \cdot \cancel{3}} = 1 \cdot 1 \cdot 1 \cdot 5 = 5$

Practice:

Reduce 1–4.

1. $\dfrac{2 \cdot 2 \cdot 3 \cdot 4 \cdot 5}{2 \cdot 2 \cdot 4 \cdot 5} = $ _____

2. $\dfrac{2 \cdot 2 \cdot 3}{2 \cdot 3 \cdot 4 \cdot 4} = $ _____

3. $\dfrac{2 \cdot 3 \cdot 3 \cdot 5}{2 \cdot 2 \cdot 3 \cdot 3 \cdot 4} = $ _____

4. $\dfrac{2 \cdot 3 \cdot 4 \cdot 4}{2 \cdot 2 \cdot 2 \cdot 3 \cdot 3 \cdot 4} = $ _____

Name _____

Reteaching 55

Math Course 1, Lesson 55

• **Common Denominators, Part 1**

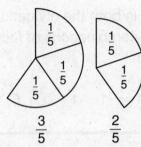

The *common denominator* is 5.

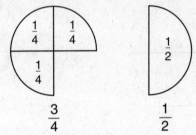

These fractions *do not* have common denominators.

To add or subtract fractions that do not have common denominators:
- Find the fraction with the largest denominator.
- Decide if that denominator can be the common denominator.
- Rename the fractions.
- Add or subtract.
- Simplify your answer.

Example:

$$\begin{array}{r} \frac{3}{4} = \frac{3}{4} \\ + \frac{1}{2} \times \frac{2}{2} = \frac{2}{4} = 1\frac{1}{4} \\ \hline \frac{5}{4} \end{array}$$

Practice:

Simplify 1–5.

1. $\begin{array}{r} \frac{2}{3} \\ + \frac{1}{6} \\ \hline \end{array}$

2. $\begin{array}{r} \frac{1}{2} \\ + \frac{1}{4} \\ \hline \end{array}$

3. $\begin{array}{r} \frac{1}{2} \\ - \frac{3}{8} \\ \hline \end{array}$

4. $\frac{3}{4} + \frac{3}{8} =$ _____

5. $\frac{7}{12} - \frac{1}{3} =$ _____

Name _____

Reteaching 56

Math Course 1, Lesson 56

• Common Denominators, Part 2

If **both** fractions must be renamed:
- Find a common denominator by multiplying the denominators.
- Rename fractions using the common denominator.
- Add or subtract. Simplify as necessary.
- Short cut: Using the LCM of the denominator for the common denominator often saves time.

Example:

$$\frac{2}{3} \times \frac{4}{4} = \frac{8}{12}$$
$$+ \frac{2}{4} \times \frac{3}{3} = \frac{6}{12} = 1\frac{2}{12} = 1\frac{1}{6}$$
$$\overline{\frac{14}{12}}$$

Renaming fractions can also help us compare fractions.

Example: $\frac{2}{3} \bigcirc \frac{3}{4} \longrightarrow \frac{2}{3} \times \frac{4}{4} = \frac{8}{12} \bigcirc \frac{3}{4} \times \frac{3}{3} = \frac{9}{12}$

$\frac{8}{12} < \frac{9}{12}$ so $\frac{2}{3} < \frac{3}{4}$

Practice:

Simplify 1–4.

1. $\frac{1}{4} + \frac{1}{5} = $ _____

2. $\frac{1}{2} + \frac{4}{5} = $ _____

3. $\frac{2}{3} - \frac{1}{2} = $ _____

4. $\frac{2}{3} - \frac{1}{4} = $ _____

Compare 5–6.

5. $\frac{1}{2} \bigcirc \frac{2}{3}$

6. $\frac{2}{3} \bigcirc \frac{6}{9}$

Saxon Math Course 1

Name _____

Reteaching 57
Math Course 1, Lesson 57

- **Adding and Subtracting Fractions: Three Steps**

When adding and subtracting fractions, remember the S. O. S. method.
- **S**hape—Write the problem in the correct shape. Rename fractions to have common denominators. (Find the LCM. Use the times table for help.)
- **O**perate—Add or subtract.
- **S**implify—Reduce or convert.

Practice:

Simplify 1–6.

1. $\frac{1}{3} + \frac{5}{9}$ = _____

2. $\frac{1}{4} + \frac{5}{6}$ = _____

3. $\frac{4}{5} + \frac{1}{2}$ = _____

4. $\frac{1}{2} - \frac{1}{3}$ = _____

5. $\frac{1}{2} - \frac{1}{8}$ = _____

6. $\frac{3}{4} - \frac{1}{3}$ = _____

Name _____

Reteaching 58

Math Course 1, Lesson 58

• **Probability and Chance**

Probability and chance **both tell** how likely an event is to happen.
  • **Probability** tells it with a (reduced) **fraction**.
  • **Chance** tells it with a **percent**.

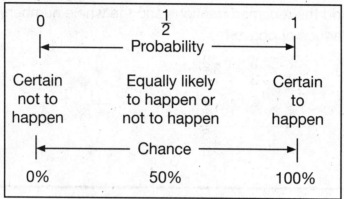

Example: a. What is the **probability** that the spinner will stop on a number greater than 1? $\frac{3}{4}$

b. What is the **chance** of spinning a number greater than 1? 75%

Example: What is the **probability** of pulling a blue marble out of the bag?

$$\frac{\text{Number of favorable outcomes}}{\text{Number of possible outcomes}} = \frac{\text{Number of blue marbles}}{\text{Total marbles}} = \frac{3}{6}$$

Then reduce. $\frac{3}{6} = \frac{1}{2}$

Practice:

1. If a number cube is rolled, what is the probability that the number rolled will be greater than 3? _____

2. What is the probability of rolling a 1 with one roll of a number cube? _____

3. A bag contains 2 red marbles, 3 blue marbles, and 5 green marbles. If one marble is drawn from the bag, what is the chance that the marble will be red? _____

4. The face of this spinner is divided into six congruent sectors. If the spinner is spun once, what is the chance that it will stop on a 2? _____

Name _____

Reteaching 59

Math Course 1, Lesson 59

• Adding Mixed Numbers

To add mixed numbers with *unlike* denominators, remember the S. O. S. method.
- **S**hape—Write the problem in the correct shape. Rename fractions to have common denominators. (Find the LCM. Use the times table for help.)
- **O**perate—Add the renamed fractions and the whole numbers.
- **S**implify—Reduce or convert.

Example:
$$3\frac{1}{2} \times \frac{3}{3} = 3\frac{3}{6}$$
$$+\ 1\frac{1}{3} \times \frac{2}{2} = 1\frac{2}{6}$$
$$\overline{4\frac{5}{6}}$$

Practice:

Simplify 1–6.

1. $1\frac{1}{2} + 3\frac{1}{3} =$ _____

2. $6\frac{3}{4} + 4\frac{3}{8} =$ _____

3. $5\frac{7}{9} + 1\frac{1}{3} =$ _____

4. $2\frac{1}{3} + 1\frac{3}{4} =$ _____

5. $2\frac{3}{8} + 3\frac{1}{2} =$ _____

6. $3\frac{1}{3} + 2\frac{1}{6} =$ _____

Reteaching 60

Math Course 1, Lesson 60

- **Polygons**

  - **Polygons** are closed, flat shapes made from straight lines.
  - Two sides of a polygon meet at a **vertex.**
  - A **quadrilateral** is a polygon with four sides.
  - The sides of a **regular** polygon are the same length.
  - The angles of a **regular** polygon are equal.
  - A **square** is a regular quadrilateral.

Polygons

Shape	Number of Sides	Name of Polygon
△	3	triangle
▭	4	quadrilateral
⬠	5	pentagon
⬡	6	hexagon
⯃	8	octagon

Practice:

1. A pentagon is a polygon with how many sides? _____

2. What is a regular quadrilateral? _____

3. What is the perimeter of a regular hexagon if each side is 6 inches long?

   _____

4. The perimeter of a regular octagon is 96 mm. How long is each side?

   _____

5. Each side of a regular hexagon is 16 cm. What is the perimeter?

   _____

Saxon Math Course 1

Name _____

Reteaching Inv.

Math Course 1, Investigation 6

• Attributes of Geometric Solids

- **Geometric solids** are three-dimensional shapes.
- They have length, width, and height.
- The volume of a solid is the amount of space it occupies.
- We can count cubes to find the volume of a solid.

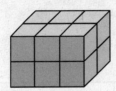

Practice:

1. A rectangular prism has how many faces?

   _____

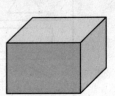

2. A pyramid with a square base has how many faces?

   _____

3. A cube has how many more edges than faces?

   _____

4. How many small cubes are used to form this larger cube?

   _____

Name _____

Reteaching 61
Math Course 1, Lesson 61

- **Adding Three or More Fractions**

- To add 3 or more fractions:
    1. Find a common denominator. Look for the least common multiple (LCM).
    2. Rename the fractions.
    3. Add whole numbers and fractions.
    4. Simplify if possible.

    Example: $\frac{1}{2} + \frac{1}{4} + \frac{3}{8}$ LCM is 8. Rename all fractions as eighths.

    $\frac{4}{8} + \frac{2}{8} + \frac{3}{8} = \frac{9}{8} = 1\frac{1}{8}$ Add and simplify.

Practice:

Simplify 1–6.

1. $\frac{1}{2} + \frac{1}{3} + \frac{5}{6} = $ _____

2. $\frac{3}{4} + \frac{3}{8} + \frac{1}{2} = $ _____

3. $\frac{1}{4} + \frac{1}{2} + \frac{2}{3} = $ _____

4. $2\frac{3}{4} + 1\frac{1}{2} + 3\frac{5}{8} = $ _____

5. $2\frac{3}{8} + 3\frac{1}{2} + 2\frac{1}{4} = $ _____

6. $2\frac{1}{2} + 2\frac{1}{6} + 2\frac{2}{3} = $ _____

Name _____

Reteaching 62

Math Course 1, Lesson 62

- **Writing Mixed Numbers as Improper Fractions**

    - To change mixed numbers to improper fractions, do one of the following:
    1. Cut the wholes into parts. **Count** the number of parts.

    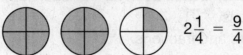

    $2\frac{1}{4} = \frac{9}{4}$

    2. Change the whole number into a fraction. Remember that $1 = \frac{4}{4}$.

    $$2\frac{1}{4} = \frac{4}{4} + \frac{4}{4} + \frac{1}{4} = \frac{9}{4}$$

    Try this shortcut:
    1. Multiply the denominator times the whole number: $4 \times 2 = 8$
    2. Add this product to the numerator: $8 + 1 = 9$
    3. Keep the original denominator: $\frac{9}{4}$

    Example: $2\frac{1}{4}$ Multiply; then add. $(4 \times 2) + 1 \rightarrow \frac{9}{4}$

Practice:

Simplify 1–4.

1. $2\frac{2}{3} =$ _____

2. $3\frac{3}{4} =$ _____

3. $1\frac{7}{8} =$ _____

4. $4\frac{5}{6} =$ _____

5. Write $3\frac{1}{3}$ as an improper fraction. Then multiply the improper fraction by $\frac{1}{4}$. Write the product as a reduced fraction.

_____

6. Write $2\frac{3}{4}$ as an improper fraction. Then multiply the improper fraction by $\frac{1}{2}$. Write the product as a reduced fraction.

_____

Name _____

Reteaching 63
Math Course 1, Lesson 63

• Subtracting Mixed Numbers with Regrouping, Part 2

- To subtract mixed numbers:
    1. **Rename** the fractions to have **common denominators.**
    2. If needed, **regroup. Combine** the renamed fractions in step 1 with the given fraction.
    3. Subtract. Simplify if possible.

Example:

Rename

$5\frac{1}{2} = 5\frac{3}{6}$
$-1\frac{2}{3} = 1\frac{4}{6}$

Regroup

$\overset{4}{\cancel{5}}\frac{3}{6} + \frac{6}{6} =$

Combine

$4\frac{9}{6}$
$-1\frac{4}{6}$
$\overline{3\frac{5}{6}}$

Practice:

Simplify 1–6.

1. $4\frac{5}{8}$
    $-1\frac{1}{2}$
    $\overline{}$

2. $3\frac{3}{4}$
    $-2\frac{5}{12}$
    $\overline{}$

3. $5\frac{1}{2} - 3\frac{2}{5} =$ _____

4. $4\frac{1}{4} - 1\frac{7}{8} =$ _____

5. $6\frac{1}{2} - 2\frac{5}{6} =$ _____

6. $2\frac{1}{4} - \frac{5}{8} =$ _____

Saxon Math Course 1

Name _____

Reteaching 64

Math Course 1, Lesson 64

- **Classifying Quadrilaterals**

  - A *square* is a special kind of rectangle.
  - A *rectangle* is a special kind of parallelogram.
  - A *parallelogram* is a special kind of quadrilateral.
  - A *quadrilateral* is a special kind of polygon.
  - A *square* is also a special kind of rhombus.

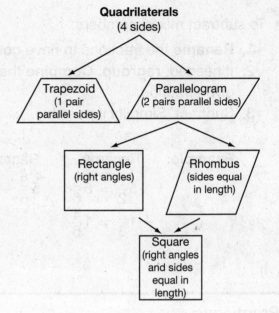

Practice:

1. Which figure is a quadrilateral? _____

   A. ○   B. ⬡   C. ▱(trapezoid)   D. △

2. Which figure is not a quadrilateral? _____

   A. ⬠   B. ▭   C. ▱   D. ▱(trapezoid)

3. Which quadrilateral is a rhombus? _____

   A.   B. □   C. ▱   D. ▭

4. Which figure is not a parallelogram? _____

   A. rectangle   B. rhombus   C. square   D. trapezoid

5. True or false: A rhombus is a special kind of rectangle. _____

Name _____

Reteaching 65
Math Course 1, Lesson 65

- **Prime Factorization**

  - A **prime number** has only two factors—itself and 1.
  - A **composite number** has more than two factors.
  - **Prime factorization** is writing a composite number as a product of its prime factors.

Division by Primes	**Factor Trees**
1. Divide by smallest prime number factor.	1. List two factors.
2. Stack divisions. Continue to divide until the quotient is 1.	2. Continue to factor until each factor is a prime number.
3. Write the factors in order.	3. Circle the prime numbers. Remember: 1 is not prime.
	4. Write the factors in order.

Example:

```
        1
     5)5
    3)15
    2)30
    2)60
```
$60 = 2 \cdot 2 \cdot 3 \cdot 5$

Example:

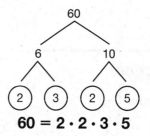

$60 = 2 \cdot 2 \cdot 3 \cdot 5$

Practice:

1. Twenty-eight is a composite number. Use division by primes to find the prime factorization of 28. _____

2. Forty-five is a composite number. Use a factor tree to find the prime factorization of 45. _____

3. Thirty-two is a composite number. Use division by primes to find the prime factorization of 32. _____

4. Fifty-four is a composite number. Use a factor tree to find the prime factorization of 54. _____

Name _____

Reteaching 66

Math Course 1, Lesson 66

- **Multiplying Mixed Numbers**

- To multiply mixed numbers:
    1. First, write the numbers in fraction form.
    2. Change the mixed numbers to improper ("top heavy") fractions.
    3. Multiply numerators and denominators.
    4. Write whole numbers as improper fractions with a denominator of 1.
    5. Simplify the product.

Example: Change mixed numbers to improper fractions first.

$$2\frac{1}{2} \times 1\frac{2}{3}$$

$$\frac{5}{2} \times \frac{5}{3} = \frac{25}{6} \qquad \frac{25}{6} = 4\frac{1}{6}$$

Multiply.   Then simplify.

Practice:

Simplify 1–6.

1. $1\frac{1}{3} \times 1\frac{1}{4} =$ _____

2. $1\frac{2}{3} \times 2\frac{1}{2} =$ _____

3. $3\frac{1}{3} \times 2 =$ _____

4. $3 \times 2\frac{2}{3} =$ _____

5. $1\frac{3}{4} \times 2\frac{1}{2} =$ _____

6. $2\frac{1}{4} \times 1\frac{1}{2} =$ _____

Name _____

Reteaching 67

Math Course 1, Lesson 67

- **Using Prime Factorization to Reduce Fractions**

- To reduce fractions using prime factorization:
  1. Write the prime factorization of the numerator and denominator.
  2. Then reduce the common factors and multiply the remaining factors.

    Example:

    1. $\dfrac{375}{1000} = \dfrac{3 \cdot 5 \cdot 5 \cdot 5}{2 \cdot 2 \cdot 2 \cdot 5 \cdot 5 \cdot 5}$

    2. $\dfrac{3 \cdot \cancel{5} \cdot \cancel{5} \cdot \cancel{5}}{2 \cdot 2 \cdot 2 \cdot \cancel{5} \cdot \cancel{5} \cdot \cancel{5}} = \dfrac{3}{8}$

Practice:

1. Write the prime factorization of the numerator and denominator of $\dfrac{16}{36}$.

    Then reduce. _____

2. Write the prime factorization of the numerator and denominator of $\dfrac{40}{72}$.

    Then reduce. _____

3. Write the prime factorization of the numerator and denominator of $\dfrac{125}{200}$.

    Then reduce. _____

4. Write the prime factorizations of 56 and 88 to reduce $\dfrac{56}{88}$. _____

5. Write the prime factorizations of 63 and 90 to reduce $\dfrac{63}{90}$. _____

6. Write the prime factorizations of 288 and 336 to reduce $\dfrac{288}{336}$. _____

Name _____

Reteaching 68

Math Course 1, Lesson 68

- **Dividing Mixed Numbers**

- To divide mixed numbers:
  1. Write the mixed numbers as improper fractions.
  2. Multiply the first fraction by the reciprocal of the second fraction. (Flip the second fraction.)
  3. Multiply numerators and denominators.
  4. Simplify the answer.

Example:

$2\frac{2}{3} \div 1\frac{1}{2}$

1. $\frac{8}{3} \div \frac{3}{2}$
2. $\frac{8}{3} \times \frac{2}{3}$
3. $\frac{8}{3} \times \frac{2}{3} = \frac{16}{9}$
4. $\frac{16}{9} = 1\frac{7}{9}$

Remember: The reciprocal of a whole number (such as 4) is $\frac{1}{\text{the number}}$ $\left(\frac{1}{4}\right)$.

Practice:

Simplify 1–6.

1. $1\frac{1}{3} \div 3 =$ _____

2. $2\frac{1}{4} \div 1\frac{1}{4} =$ _____

3. $3 \div 1\frac{1}{2} =$ _____

4. $4 \div 1\frac{5}{7} =$ _____

5. $2\frac{1}{2} \div 1\frac{2}{5} =$ _____

6. $2\frac{2}{3} \div 4 =$ _____

Reteaching 69

Math Course 1, Lesson 69

- **Lengths of Segments**
- **Complementary and Supplementary Angles**

Lengths of Segments

In this figure, the length of segment *JK* is 3 cm and the length of segment *JL* is 5 cm. What is the length of segment *KL*?

The length of segment *JK* plus the length of segment *KL* equals the length of segment *JL*.
3 cm + l = 5 cm
l = 2 cm   So, the length of segment *KL* is 2 cm.

Complementary and Supplementary Angles

Complementary angles are two angles whose measures total 90°.
Supplementary angles are two angles whose measures total 180°.
∠ABC and ∠CBD are complementary
∠ABD and ∠DBE are supplementary

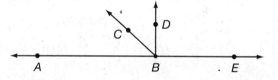

Practice:

1. In this figure, the length of segment *OP* is 6 cm and the length of segment *NP* is 10 cm. Find the length of segment *NO*. _____

2. A complement of a 30° angle is an angle that measures how many degrees?

_____

3. A supplement of a 70° angle is an angle that measures how many degrees?

_____

4. Name two angles in the figure at right that appear to be supplementary.

_____

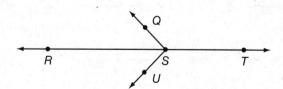

5. Name two angles in the figure at right that appear to be complementary.

_____

Name _____

Reteaching 7
Math Course 1, Lesson 70

- **Reducing Fractions Before Multiplying**

  - Reducing before multiplying is also known as **canceling.**
  - Canceling may be done to the terms of **multiplied** fractions only.
  - Look for common terms in a diagonal.
  - Reduce the common terms by dividing by a common factor.

    Example: $\frac{10}{9} \times \frac{6}{5}$

    Divide 10 and 5 by 5.

    Divide 9 and 6 by 3.

    Multiply the remaining terms. $\frac{2}{3} \times \frac{2}{1} = \frac{4}{3}$

    Reduce. $\frac{4}{3} = 1\frac{1}{3}$

  - Reducing before you multiply can save you from reducing after you multiply.

    Long Way **Short Way**

    $\frac{3}{5} \times \frac{2}{3} = \frac{6}{15}$ $\frac{6}{15}$ reduces to $\frac{2}{5}$ $\frac{\cancel{3}^1}{5} \times \frac{2}{\cancel{3}_1} = \frac{2}{5}$

Practice:

Simplify 1–6.

1. $1\frac{2}{3} \times 1\frac{1}{2} =$ _____

2. $2\frac{1}{2} \times 2\frac{2}{3} =$ _____

3. $3\frac{1}{3} \times 1\frac{4}{5} =$ _____

4. $2\frac{2}{3} \times 1\frac{1}{8} =$ _____

5. $1\frac{2}{9} \times 3 =$ _____

6. $4 \times 2\frac{3}{4} =$ _____

Name _____

Reteaching Inv. 7

Math Course 1, Investigation 7

• **The Coordinate Plane**

Every point on a coordinate plane is named with two numbers. The first number shows how far left or right; the second numbers shows how far up or down.

Practice:

Refer to the coordinate plane on the right for problems 1 and 2.

1. Which point has the coordinates (3, −1)?

   _____

2. What are the coordinates of point *D*?

   _____

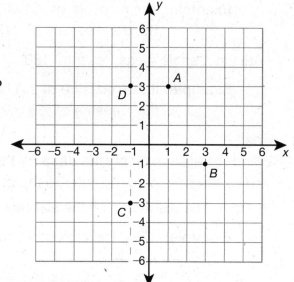

The coordinates of the four vertices of a rectangle are (1, −1), (4, −1), (4, 4), and (1, 4). Use this information for problems 3 and 4.

3. The perimeter of the rectangle is how many units? _____

4. How many square units is the area of the rectangle? _____

5. What are the coordinates of the point halfway between (−1, 1) and (3, 1)?

   _____

6. The coordinates of three vertices of a square are (2, 2), (−1, 2), and (−1, −1). What are the coordinates of the fourth vertex?

   _____

Saxon Math Course 1   © Harcourt Achieve Inc. and Stephen Hake. All rights reserved.

Name _____

Reteaching 7

Math Course 1, Lesson 71

• Parallelograms

Angle properties:
- The four angle measures total 360°
- Opposite angles have equal measures.

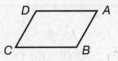

   Example: The measure of ∠A is 60°.
   So, the measure of ∠C is 60°.

- Adjacent angle measures total 180°. (Adjacent angles share a side.)

   Example: If the measure of ∠A is 60°,
   the measure of ∠B must be 120°.

Area:
To find the **area**, multiply the **base** by the **height**.
- The height is perpendicular to the base.
- Do not be distracted by the slanted side.

Example: Find the area of this parallelogram.

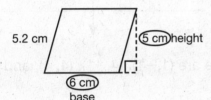

$A = bh$
$A = (6)(5)$
$A = 30$ sq. cm

Practice:

1. What is the area of this parallelogram? _____

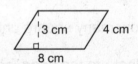

Refer to parallelogram *ABCD* to answer problems 2–5.

2. What is the perimeter of the parallelogram? _____

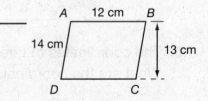

3. What is the area of the parallelogram? _____

4. If angle *A* measures 100°, what is the measure of angle *C*? _____

5. If angle *A* measures 100°, what is the measure of angle *B*? _____

Name _____

Reteaching 72
Math Course 1, Lesson 72

- **Fractions Chart**
- **Multiplying Three Fractions**

- To use the S.O.S. method in the Fractions Chart:
  1. Write the problem in the correct *shape*.
  2. Perform the *operation*.
  3. *Simplify* the answer.

- To multiply three or more fractions:
  Step 1: Write the numbers in fraction form.
  Step 2: Cancel terms by reducing numerator-denominator pairs with common factors. Multiply the remaining terms.
  Step 3: Simplify if possible.

Fractions Chart

	+ −	× ÷
Shape	Write fractions with common denominators.	Write numbers in fraction form.
Operate	Add or subtract the numerators.	**×** cancel. $\frac{n \times n}{d \times d}$ / **÷** Find reciprocal of divisor; then
Simplify	Reduce fractions. Convert improper fractions.	

Example: 1. $\frac{2}{3} \times \frac{8}{5} \times \frac{3}{4}$;

2. $\frac{2}{\cancel{3}_1} \times \frac{\cancel{8}^2}{5} \times \frac{\cancel{3}^1}{\cancel{4}_1}$; $\frac{2}{1} \times \frac{2}{5} \times \frac{1}{1} = \frac{4}{5}$;

3. $\frac{4}{5}$ is reduced to lowest terms.

Practice:

1. What is the first step in adding fractions? _____

2. What is the last step in multiplying fractions?
   _____

Simplify 3–6.

3. $\frac{1}{2} \times \frac{3}{4} \times \frac{4}{9} =$ _____

4. $\frac{5}{6} \times \frac{3}{5} \times \frac{2}{3} =$ _____

5. $\frac{5}{7} \times \frac{3}{10} \times \frac{2}{3} =$ _____

6. $\frac{3}{8} \times \frac{2}{3} \times \frac{3}{4} =$ _____

Name _____

Reteaching 73
Math Course 1, Lesson 73

- **Exponents**
- **Writing Decimal Numbers as Fractions, Part 2**

  - The **exponent** shows how many times the base number is used as a factor.
  - We read numbers with exponents as **powers**.
    The exponent 2 is read "squared."
    The exponent 3 is read "cubed."

    Examples: $5^2 = 5 \times 5 = 25$

    (five squared)

    $3^4 = 3 \times 3 \times 3 \times 3 = 81$

    (three to the fourth power)

  - Using a factor tree or division by primes, we find the prime factorization of 1000.
    $$1000 = 2 \cdot 2 \cdot 2 \cdot 5 \cdot 5 \cdot 5$$

  We group the three 2's and the three 5's with exponents.
  $$1000 = 2^3 \cdot 5^3$$

  - To write decimal numbers as fractions or mixed numbers:
    1. Write the digits after the decimal point as the numerator.
    2. Write the denominator (10 or 100 or 1000 ...) indicated by the number of decimal places in the decimal number.
    3. Any digits to the left of the decimal point represent the whole number.
    4. Reduce the fraction if possible.

    Examples: $0.5 = \frac{5}{10} = \frac{1}{2}$

    $3.75 = 3\frac{75}{100} = 3\frac{3}{4}$

Practice:

1. Write the prime factorization of 36 using exponents. _____

2. Write the prime factorization of 200 using exponents. _____

Write each decimal number as a fraction or mixed number for 3–6.

3. $0.53 =$ _____   4. $8.5 =$ _____

5. $0.4 =$ _____   6. $1.25 =$ _____

Name _____

Reteaching 74
Math Course 1, Lesson 74

- **Writing Fractions as Decimal Numbers**

To write a fraction as a decimal:
- Divide the numerator by the **denominator.**
- Keep the whole number.

Examples: $\frac{3}{4} \longrightarrow 4\overline{)3.00}^{\,0.75}$

$2\frac{2}{5} \longrightarrow 5\overline{)2.0}^{\,0.4} \longrightarrow 2.4$

Practice:

1. Write $\frac{3}{5}$ in decimal form. _____

2. Write $\frac{3}{8}$ in decimal form. _____

3. Write $2\frac{1}{4}$ as a decimal numeral. Then add it to 3.5 and find the sum.

   _____

4. Write $1\frac{3}{4}$ as a decimal numeral. Then add it to 4.75 and find the sum.

   _____

Simplify 5–6.

5. $5.25 - 2\frac{1}{2} =$ _____

6. $4.5 + \frac{3}{10} =$ _____

Name _____   **Reteaching 75**
Math Course 1, Lesson 75

- **Writing Fractions and Decimals as Percents, Part 1**

  - A percent is a fraction with a denominator of 100.
    Instead of writing the denominator, we write a percent sign.
    Example: $\frac{25}{100} = 25\%$

  - To write a fraction as a percent, first write an equivalent fraction that has a denominator of 100.
    Example: $\frac{3}{10} \cdot \frac{10}{10} = \frac{30}{100} = 30\%$

  - To change a decimal to a percent:
    1. Write as a fraction with a denominator of 100.
       Example: $0.8 = 0.80 = \frac{80}{100} = 80\%$
    2. Shift the decimal point two places to the right.
       Example: $0.25 \longrightarrow 0.25 \longrightarrow 25\%$

Practice:

1. Write $\frac{3}{10}$ as a percent. _____

2. Write $\frac{2}{5}$ as a percent. _____

3. Write 0.06 as a percent. _____

4. Carlos correctly answer 46 of the 50 questions.

   What percent of the questions did he answer correctly? _____

5. Victoria correctly answered 17 of the 20 questions.

   What percent of the questions did she answer correctly? _____

6. What percent of this circle is shaded? _____

Name _____

Reteaching 76
Math Course 1, Lesson 76

- **Comparing Fractions by Converting to Decimal Form**

  - Another way to compare fractions:
    Convert the fractions to decimal form.

    Example: $\frac{3}{5} \bigcirc \frac{5}{8} \longrightarrow 0.6 \bigcirc 0.625$

    Since 0.6 is less than 0.625, we know that $\frac{3}{5}$ is less than $\frac{5}{8}$.

    $$\frac{3}{5} < \frac{5}{8}$$

  - If a problem contains fractions and decimal numbers, convert to the same form.

    Example: $0.7 \bigcirc \frac{3}{4} \longrightarrow 0.7 \, \text{\textcircled{<}} \, 0.75$

Practice:

Compare 1–6.

1. $\frac{3}{8} \bigcirc 0.4$

2. $\frac{1}{2} \bigcirc 0.5$

3. $0.6 \bigcirc \frac{2}{5}$

4. $\frac{4}{5} \bigcirc \frac{7}{8}$

5. $\frac{7}{4} \bigcirc \frac{7}{10}$

6. $\frac{1}{2} \bigcirc \frac{3}{5}$

Name _____

Reteaching 77

Math Course 1, Lesson 77

- **Finding Unstated Information in Fraction Problems**

The following sentence directly states information about the number of boys in the class.

It *indirectly* states information about the number of girls in the class.

    Three fourths of the 28 students in the class are boys.

Diagram the statement:
- Into how many parts is the class divided?   4 parts
- How many are in each part?   28 ÷ 4 = 7
- How many parts are boys?   3 parts
- How many boys are in the class?   3 × 7 = 21 boys
- How many parts are girls?   1 part
- How many girls are in the class?   7 girls

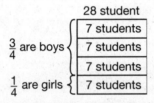

Practice:

1. If $\frac{1}{4}$ of the dozen eggs were cracked, how many eggs were not cracked? _____

2. If $\frac{2}{5}$ of the 200 flowers bloomed, how many flowers did not bloom? _____

3. Two thirds of the 24 runners finished. How many did not finish? _____

4. Three eighths of the 32 students are in chorus.

   How many students are not in chorus? _____

Name _____

Reteaching 78
Math Course 1, Lesson 78

• **Capacity**

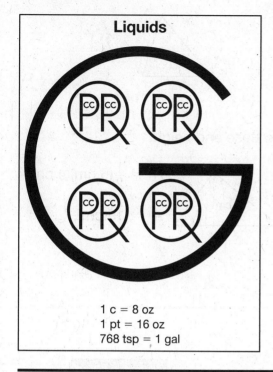

1 gallon   ½ gallon   1 quart   1 pint   1 cup

Equivalence Table for Units of Liquid Measure

U.S. Customary System	Metric System
1 gallon = 4 quarts	
1 quart = 2 pints	
1 pint = 2 cups	1 liter = 1000 mililiters
1 pint = 16 ounces	
1 cup = 8 ounces	

1 c = 8 oz
1 pt = 16 oz
768 tsp = 1 gal

Practice:

1. A pint is what fraction of a quart? _____

2. A quart of milk is how many cups? _____

3. How many quarts are in $\frac{1}{2}$ gallon? _____

4. A gallon of juice will fill how many 8-ounce cups? _____

Name _____

Reteaching 79

Math Course 1, Lesson 79

- **Area of a Triangle**

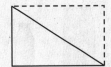  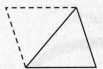

- Notice that the area of any triangle is:
  $\frac{1}{2}$ the area of a parallelogram with the same base and height

- So the formula for the area of a triangle is:

  $A = \frac{1}{2}bh$ or $A = \frac{bh}{2}$

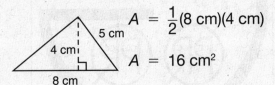

- The height is perpendicular to the base.
- Area is expressed in **square** units (16 cm²).

Practice:

1. Figure ABCD is a rectangle. Segment AD is 5 cm long, segment DC is 4 cm long, and segment AC is 6 cm long. What is the area of triangle ADC?

   _____

   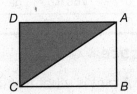

2. What is the area of this triangle? _____

   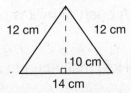

3. What is the area of this triangle? _____

   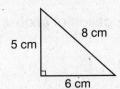

4. What is the area of the shaded part of this parallelogram? _____

   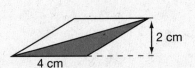

5. The area of the shaded part of this rectangle is half of the area of the rectangle.

   What is the area of the shaded triangle? _____

   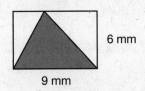

Reteaching 80
Math Course 1, Lesson 80

- **Finding Scale Factor to Solve Ratio Problems**

- Use a ratio box to sort the numbers in a ratio problem.
- Ratio numbers and actual counts are related by a **scale factor**.
- Find the factor by which the actual count was reduced to form the ratio.

Example: The ratio of boys to girls in the class was 3 to 2. If there were 8 girls in the class, how many boys were there?

	Ratio	Actual Count
Boys	3	
Girls	2	8

Ratio × scale factor = actual count

2 × 4 = 8

3 × 4 = 12

There were 12 boys in the class.

Practice:

1. The ratio of red flowers to white flowers in the garden was 4 to 3.

   If there were 30 white flowers, how many red flowers were there? _____

2. The ratio of girls to boys on the soccer team was 3 to 5.

   If there were 10 boys on the team, how many girls were there? _____

3. The ratio of vans to cars in the parking lot was 2 to 4.

   If there were 24 cars in the lot, how many vans were there? _____

4. The ratio of white chalk to colored chalk in the box was 6 to 4. If there were 16 pieces of colored chalk in the box, how many pieces of white chalk were there?

   _____

Name _____

Reteaching Inv.

Math Course 1, Investigation 8

- **Geometric Construction of Bisectors**

  - To construct a perpendicular bisector:
    1. Set the compass on each end point and swing an arc above and below the segment.
    2. Draw a line through the points where the arcs intersect.

  - To construct an angle bisector:
    1. Set the compass on the vertex and swing an arc across the angle.
    2. Set the compass on each new intersection and swing an arc.
    3. Draw a ray from the vertex through the intersection of the arcs.

Practice:

Draw a perpendicular bisector of each segment.

1. D•——————————•E

2. S•————•T

Bisect each angle.

3.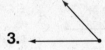

4.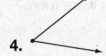

Name _____

Reteaching 81
Math Course 1, Lesson 81

- **Arithmetic with Units of Measure**

  - If units are not the same, convert first.

    Example: 2 ft + 12 in. ⟶ 24 in. + 12 in. or 2 ft + 1 ft

  - To add or subtract measures, keep the unit.

    Example: 24 in. + 12 in. = 36 in. or 2 ft + 1 ft = 3 ft

  - To multiply measures, multiply the units.

    Example: 2 cm × 3 cm = $\underbrace{2 \cdot 3}_{6}$ $\underbrace{cm \cdot cm}_{cm^2}$

  - To divide measures, divide the units.

    Example: $\dfrac{21\ cm^2}{7\ cm} = \dfrac{\overset{3}{\cancel{21}}}{\underset{1}{\cancel{7}}} \dfrac{\cancel{cm} \cdot cm}{\cancel{cm}} = 3\ cm$

  - Some units will not reduce.

    Example: $\dfrac{300\ mi}{6\ hr} = \dfrac{\overset{50}{\cancel{300}}}{\underset{1}{\cancel{6}}} \dfrac{mi}{hr} = 50\ \dfrac{mi}{hr}$

Practice:

Simplify 1–6.

1. 3 ft − 4 in. = (Write the difference in inches.) _____

2. 2 ft − 10 in. = (Write the difference in inches.) _____

3. 4 ft + 6 in. = (Write the sum in inches.) _____

4. 3 ft × 6 ft = _____

5. $\dfrac{25\ cm^2}{5\ cm}$ = _____

6. $\dfrac{500\ mi}{10\ hr}$ = _____

Reteaching 82

Math Course 1, Lesson 82

- **Volume of a Rectangular Prism**

  - The **volume** of a shape is the amount of space the shape occupies. Volume is measured in **cubic width**.

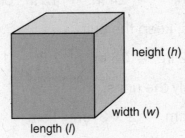

  - $V = lwh$ (volume = length × width × height)

    Example: What is the volume of this rectangular prism?
    $V = lwh$
    $V = (5)(3)(4)$
    $V = 60$ in.^{3}
    (in.^{3} means cubic inches)

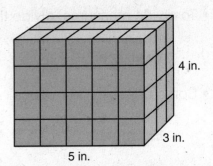

Practice:

1. What is the volume of a shoe box that is 11 inches long, 6 inches wide, and 4 inches high? _____

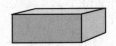

2. What is the volume of this rectangular prism?

   _____

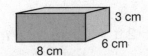

3. What is the volume of a cube that has edges 5 inches long? _____

4. How many 1-centimeter sugar cubes would be needed to form this rectangular prism? _____

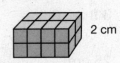

5. What is the volume of this rectangular prism?

   _____

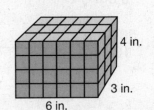

Name _____

Reteaching 83
Math Course 1, Lesson 83

• **Proportions**

A **proportion** is a true statement that two ratios are equal.

$$\frac{3}{4} = \frac{6}{8}$$ Three is to four as six is to eight.

- To find a missing term in a proportion, find the number that is multiplied by the first term to get the second term.
- Whatever has been done to the numerator, do to the denominator (and vice versa).

Example: $\frac{3}{5} = \frac{6}{a}$ $3 \times 2 = 6$
so $5 \times 2 = 10$
$a = 10$ $\frac{3}{5} \times \frac{2}{2} = \frac{6}{10}$

Practice:

1. Which ratio forms a proportion with $\frac{4}{5}$? _____

   A. $\frac{8}{9}$ B. $\frac{16}{20}$ C. $\frac{14}{15}$ D. $\frac{16}{25}$

2. Write and complete this proportion:
   Three is to seven as nine is to what number?

   _____

3. Write and complete this proportion:
   Nine is to five as eighteen is to what number?

   _____

4. Write and complete this proportion:
   Six is to five as what number is to twenty-five?

   _____

Saxon Math Course 1

Name _____

Reteaching 84

Math Course 1, Lesson 84

• **Order of Operations, Part 2**

Order of Operations
1. Parentheses
2. Multiply and divide, in order, left to right.
3. Add and subtract, in order, left to right.

Example: $2(8 + 6) + 15 \div 5$ original problem
              $\underline{2(14)} + \underline{15 \div 5}$ simplified parentheses
              $28 \;+\; 3$ multiplied and divided
              31 added

Practice:

Simplify 1–6.

1. $3 \times 3 + 4 \times 5$ = _____

2. $6 \times 5 - 7 \times 2$ = _____

3. $2 + 8 \times 2 - 5$ = _____

4. $10 + 9 \div 3 - 6$ = _____

5. $6 \times 4 + 3 \times 2$ = _____

6. $3 \times (3 + 4) \div 4$ = _____

Name _____

Reteaching 85
Math Course 1, Lesson 85

- **Using Cross Products to Solve Proportions**

  - Equal fractions have equal cross products.

    $$\frac{3}{4} \times \frac{6}{8}$$

    $8 \times 3 = $ **24** $4 \times 6 = $ **24**

  - Another way to complete proportions:
    1. Cross-multiply.
    2. Divide by known factor.

    Example: $\frac{3}{5} = \frac{6}{w}$ $(5 \times 6) \div 3 = 10$ or $\frac{5 \cdot \overset{2}{\cancel{6}}}{\underset{1}{\cancel{3}}} = 10$

    Cancel.

    Example: $\frac{15}{21} = \frac{w}{70}$ $\frac{\overset{5}{\cancel{15}} \cdot \overset{10}{\cancel{70}}}{\underset{1}{\underset{7}{\cancel{21}}}} = w$ $w = 50$

Practice:

1. Complete this proportion:

   $\frac{5}{8} = \frac{10}{n}$

   $n = $ _____

2. Complete this proportion:

   $\frac{7}{k} = \frac{2}{10}$

   $k = $ _____

Solve 3–4.

3. $\frac{2}{6} = \frac{9}{w}$

   $w = $ _____

4. $\frac{4}{20} = \frac{x}{100}$

   $x = $ _____

5. Solve this proportion:

   $\frac{8}{10} = \frac{y}{35}$

   $y = $ _____

6. Solve the proportion:

   $\frac{m}{15} = \frac{10}{25}$

   $m = $ _____

Name _____

Reteaching 86
Math Course 1, Lesson 86

- **Area of a Circle**

  - The radius of a circle is 3 cm. What is the area of the circle?
    1. Find the area of a square whose sides equal the radius.
    2. Multiply that area by 3.14.

Area of square: 3 cm × 3 cm = 9 cm²
Area of circle: (3.14)(9 cm²) = 28.26 cm²
$A = \pi \cdot r^2$
Remember: radius = $\frac{1}{2}$ of the diameter
$\pi = 3.14$

Practice:

1. What is the area of the circle? _____

2. If the radius of a circle is 4 cm, what is its area? (Use 3.14 for π) _____

3. If the radius of a circle is 2 cm, what is its area? (Use 3.14 for π) _____

4. The diameter of a circle target is 14 inches.

   What is the area of the target? (Use 3.14 for π) _____

5. The diameter of a circular tray is 24 inches.

   What is the area of the tray? (Use 3.14 for π) _____

Name _____

Reteaching 87

Math Course 1, Lesson 87

- **Finding Unknown Factors**

- To find a missing factor, **divide** the product by the known factor.
  Write answers as mixed numbers (unless there are decimals in the problem).

  Examples: $5n = 21$ $\quad \begin{array}{r} 4\frac{1}{5} \\ 5\overline{)21} \end{array}$

  $n = 4\frac{1}{5}$

  $0.6m = 0.048 \quad\quad \begin{array}{r} 0.08 \\ 0.6\overline{)0.0.48} \end{array}$

  $m = 0.08$

Practice:

Solve 1–6.

1. $6n = 0.84$

   $n = $ _____

2. $4y = 7$

   $y = $ _____

3. $8f = 36$

   $f = $ _____

4. $5.5t = 22$

   $t = $ _____

5. $5m = 0.95$

   $m = $ _____

6. $1.2 = 0.2k$

   $k = $ _____

Name _____

Reteaching 88

Math Course 1, Lesson 88

- **Using Proportions to Solve Ratio Problems**

- Proportions can be used to solve many types of word problems. Use a ratio table to organize the numbers.

    Example: The ratio of salamanders to frogs was 5 to 7. If there were 20 salamanders, how many frogs were there?

	Ratio	Actual Count
Salamanders	5	20
Frogs	7	f

- Two ways to solve:

    1. Multiply by a fractional name for 1.

    $$\frac{5}{7} = \frac{20}{f} \longrightarrow \frac{5}{7} \times \frac{4}{4} = \frac{20}{28} \longrightarrow 28 \text{ frogs}$$

    2. Cross-multiply. Then divide by known factor.

    $$\frac{5}{7} = \frac{20}{f} \longrightarrow (7 \times 20) \div 5 = 28 \quad \text{or} \quad \frac{7 \cdot \overset{4}{\cancel{20}}}{\underset{1}{\cancel{5}}} = 28 \text{ frogs}$$

Practice:

1. The ratio of boys to girls on the team was 3 to 2.

    If there were 9 boys, how many girls were there? _____

2. The ratio of boys to girls in the class was 4 to 5.

    If there were 10 girls, how many boys were there? _____

3. The ratio of A's to B's on the last test was 5 to 3.

    If there were 10 A's, how many B's were there? _____

4. The ratio of goldfish to angelfish in the fish tank was 6 to 4.

    If there were 16 angelfish, how many goldfish were there? _____

Name _____

Reteaching 89
Math Course 1, Lesson 89

• Estimating Square Roots

- To find the square root of a perfect square greater than 100, use the "Guess and Check" method. Think of square roots that you know.

    Example: Simplify: $\sqrt{400}$

    Think: $\sqrt{4} = 2$ and $\sqrt{100} = 10$. So, try 20.
    $20 \times 20 = 400$
    $\sqrt{400} = 20$

- To estimate the square root of numbers that are **not** perfect squares, find the perfect square on either side of the given number.

    Example: Between which two consecutive whole numbers is $\sqrt{20}$?
    $\sqrt{20}$ is between 4($\sqrt{16}$) and 5($\sqrt{25}$)

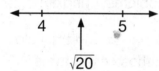

- **Irrational numbers** cannot be exactly expressed as fractions or decimals.
    $\sqrt{20} \approx 4.5$ The wavy equals sign means "approximately equal to."

Practice:

1. Between which 2 numbers is $\sqrt{14}$? _____

    A. 2 and 3   B. 3 and 4   C. 4 and 5   D. 13 and 15

2. Between which 2 numbers is $\sqrt{27}$? _____

    A. 3 and 4   B. 4 and 5   C. 5 and 6   D. 26 and 28

Simplify 3–6.

3. $\sqrt{2500}$ = _____

4. $\sqrt{3600}$ = _____

5. $\sqrt{6400}$ = _____

6. $\sqrt{196}$ = _____

Name _____

Reteaching 9
Math Course 1, Lesson 90

- **Measuring Turns**

Turns can be measured in degrees.
- A full turn is 360°.
- A half turn is 180°.
- A quarter turn is 90°.

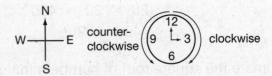

Direction can be described as:
- left or right.
- clockwise or counterclockwise.
- north, south, east, or west.

Practice:

Use the figure on the right for Problems 1 and 2.

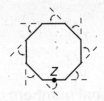

1. Eddie began at point Z and ran around the field making the turns shown. In those eight turns, Eddie turned a total of how many degrees?

   _____

2. What was the average number of degrees in each of Eddie's turns? _____

3. Angelina was heading north. She made a half turn clockwise.

   What direction was Angelina now heading? _____

4. Ryan rode his bike east. He turned 90° to the right and rode some more. Then he turned 90° to the right and rode some more.

   What direction was Ryan now heading? _____

Name _____

Reteaching Inv. 9
Math Course 1, Investigation 9

- **Experimental Probability**

- We can calculate **experimental probability** by performing an experiment repeatedly and collecting data about the experiment's outcomes.
- A survey is one type of probability experiment.
- A probability expressed as a percent is called **chance**.

Practice:

1. A bag contains red, green and blue marbles. In an experiment, you draw one marble from the bag and record its color in a list. You repeat the experiment 10 times with the following results:
   blue, green, green, green, red, green, green, blue, blue, blue.

   a. What is the probability that a marble from the bag will be blue? _____

   b. What is the probability that a marble from the bag will be green? _____

   c. What is the chance that a marble from the bag will be red? _____

2. Jan determined that the theoretical probability of tossing a coin that lands on heads is $\frac{1}{2}$. To test her calculation, Jan tossed the same coin 100 times. It landed on heads 52 times.

   a. What is the experimental probability of tossing a coin that lands on heads based on Jan's data? _____

   b. Did Jan's experiment agree with the theoretical probability that she calculated? Why or why not?

   _____

   _____

3. To plan for a surprise ice-cream sundae party, the cafeteria staff of a school conducts a survey of 30 students to find out their favorite ice cream flavors. They learn that 11 students like vanilla, 17 students like chocolate, and the rest like strawberry.

   a. What is the chance that a student in the school likes strawberry-flavored ice cream? _____

   b. If there are 720 students in the school, how many will want chocolate ice cream? _____

Name _____

Reteaching 9
Math Course 1, Lesson 91

- **Geometric Formulas**

Shape	Perimeter	Area
Square	$P = 4s$	$A = s^2$
Rectangle	$P = 2l + 2w$	$A = lw$
Parallelogram	$P = 2b + 2s$	$A = bh$
Triangle	$P = s_1 + s_2 + s_3$	$A = \frac{1}{2}bh$

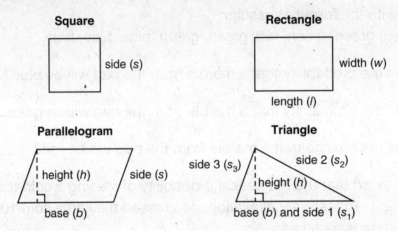

Practice:

1. Write the formula for the perimeter of a square. Then substitute 8 inches for the side. Solve the equation to find the perimeter of the square.

   _____

2. Write the formula for the area of a rectangle. Then substitute 4 cm for the length and 6 cm for the width. Solve the equation to find the area of the rectangle.

   _____

3. Write the formula for the perimeter of a parallelogram. Then substitute 3 inches for the base and 5 inches for the side. Solve the equation to find the perimeter of the parallelogram.

   _____

4. Write the formula for the area of a triangle. Then substitute 5 cm for the base and 8 cm for the height. Solve the equation to find the area of the triangle.

   _____

© Harcourt Achieve Inc. and Stephen Hake. All rights reserved.

Saxon Math Course 1

Name _____

Reteaching 92
Math Course 1, Lesson 92

- **Expanded Notation with Exponents**
- **Order of Operations with Exponents**
- **Powers of Fractions**

- To write numbers in expanded notation, we may also show whole number place values with powers of 10.

    Notice the exponent and the number of zeros it takes.

    $10^4 = 10{,}000 \quad 10^3 = 1000 \quad 10^2 = 100 \quad 10^1 = 10 \quad 10^0 = 1$

    Example: Show 186,000 in expanded notation using exponents.
    186,000
    $(1 \times 100{,}000) + (8 \times 10{,}000) + (6 \times 1000)$
    $(1 \times 10^5) \quad\quad + (8 \times 10^4) + (6 \times 10^3)$

    The exponent after the 10 is equal to the number of zeros to the right of the number.

- In the order of operations, simplify expressions before multiplying or dividing.

    1. Simplify parentheses.
    2. Simplify exponents (powers) and roots.
    3. Multiply and divide left to right.
    4. Add and subtract left to right.

    > Some students remember the order of operations by memorizing this phrase:
    > **P**lease—P is for parentheses.
    > **E**xcuse—E is for exponents.
    > **M**y **D**ear—M is for multiplication;
    > D is for division.
    > **A**unt **S**ally—A is for addition;
    > S is for subtraction.

- Exponents may be used with fractions and with decimals.
    Convert a mixed number to an improper fraction before you multiply.

    Example: $\left(1\tfrac{1}{2}\right)^2 \longrightarrow \left(\tfrac{3}{2}\right)^2 \quad\quad \tfrac{3}{2} \times \tfrac{3}{2} = \tfrac{9}{4} = 2\tfrac{1}{4}$

Practice:

Simplify 1–5.

1. $9 + 3 \times 5 - 4^2 =$ _____

2. $\left(1\tfrac{1}{3}\right)^2 =$ _____

3. $3^2 - \sqrt{9} =$ _____

4. $2^3 + \sqrt{16} - 2 \times 5 =$ _____

5. $3^3 - 2^2 + 9 \times 4 =$ _____

6. Write the standard notation: $(3 \times 10^3) + (5 \times 10^2) =$ _____

Name _____

Reteaching 93
Math Course 1, Lesson 93

• Classifying Triangles

Classifying Triangles by Sides

Characteristic	Type	Example
Three sides of equal length	Equilateral triangle	△
Two sides of equal length	Isosceles triangle	▷
Three sides of unequal length	Scalene triangle	△

Classifying Triangles by Angles

Characteristic	Type	Example
All acute angles	Acute triangle	△
One right angle	Right triangle	◣
One obtuse angle	Obtuse triangle	△

Practice:

1. Which of these terms describes triangle *ABC*? _____

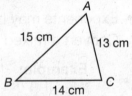

   A. acute triangle
   B. isosceles triangle
   C. right triangle
   D. obtuse triangle

2. What is the perimeter of an equilateral triangle if one of its sides is 12 inches long? _____

3. If the perimeter of an equilateral triangle is 27 inches, how long is each side? _____

4. An equilateral triangle is also what kind of triangle? _____

5. A right triangle can also be an isosceles triangle? True or false? _____

Name _____

Reteaching 94

Math Course 1, Lesson 94

- **Writing Fractions and Decimals as Percents, Part 2**

- To change a number to a percent:
    1. Multiply the number by 100%.
    2. With fractions, cancel if possible.

    Example: Write $\frac{6}{5}$ as a percent.

    $$\frac{6}{\underset{1}{\cancel{5}}} \times \frac{\overset{20}{\cancel{100}}\%}{1} = 120\%$$

    Write 1.2 as a percent.

    $$1.2 \times 100\% = 120\%$$

    3. If a fraction will not cancel to a 1 in the denominator:
        Multiply across.
        Divide the fraction.

    Example: Change $\frac{1}{3}$ to a percent.

    $$\frac{1}{3} \times \frac{100\%}{1} = \frac{100\%}{3} \longrightarrow 3\overline{)100\%}^{\,33\frac{1}{3}\%}$$

Remember: to change a fraction to a percent, multiply the fraction by 100%.
to change a decimal number to a percent, multiply by 100%.

Practice:

1. Change $\frac{2}{5}$ to a percent. _____

2. Change $\frac{1}{7}$ to a percent. _____

3. Change $\frac{3}{8}$ to a percent. _____

4. Change 0.92 to a percent. _____

5. Change 0.406 to a percent. _____

Name _____

Reteaching 95

Math Course 1, Lesson 95

- **Reducing Units Before Multiplying**

- We cancel **numbers** in fractions before multiplying.
- Also cancel **units** in measures before multiplying.

    Example: Multiply 4 miles per hour by two hours.

        Write 4 miles per hour as the ratio 4 miles over 1 hour.
        "Per" indicates division.
        Write two hours as the ratio 2 hours over 1.

$$\frac{4 \text{ miles}}{1 \text{ hour}} \times \frac{2 \text{ hours}}{1} = 8 \text{ miles}$$

Practice:

Simplify 1–3.

1. $\dfrac{8 \text{ dollar}}{1 \text{ hour}} \times 7 \text{ hours} =$ _____

2. $\dfrac{7 \text{ cents}}{1 \text{ minute}} \times 45 \text{ minutes} =$ _____

3. $\dfrac{300 \text{ miles}}{1 \text{ day}} \times 2 \text{ days} =$ _____

4. Multiply 15 teachers by 18 students per teacher. _____

5. Multiply 3.9 meters by 100 centimeters per meter. _____

Name _____

Reteaching 96

Math Course 1, Lesson 96

- **Functions**

A **function** pairs one unknown with another unknown.

1. Study the table to find the function rule.

   Example:

Position	First	Second	Third	Fourth	Fifth	Sixth
n	1	2	3	4	5	6
Term	1	4	9	16		

   What do you do to 1 to get 1? Multiply by 1 or add 1.
   What do you do to 2 to get 4? Multiply by 2 or add 2.
   What do you do to 3 to get 9? Multiply by 3 or add 6.
   What do you do to 4 to get 16? Multiply by 4 or add 12.
   What rule can apply to all the numbers?
     Each number is multiplied by itself to get the term.
   How can we generalize the rule for this sequence?
     Multiply n times itself or n^2.

2. Apply the rule of the function to find the missing numbers.
   $$5 \times 5 = 25 \qquad 6 \times 6 = 36$$
   Using the rule, you can predict what the tenth term will be.
   $$10 \times 10 \text{ or } 10^2 = 100$$

Practice:

Find the missing numbers in each function table.

1.

n	2	3	4	5	6
Term	6	9	12		

2.

n	5	10	15	20	25
Term	1	6	11		

3.

Chair	1	2	3	4	5
Legs	4	8	12		

4.

Gloves	2	4	6	8	10
Fingers	10	20	30		

Name _____

Reteaching 9
Math Course 1, Lesson 97

- **Transversals**

  - A line that intersects two or more other lines is a **transversal**.
  - ∠1 and ∠5 are **corresponding angles** (same relative position).
  - Angles between the parallel lines are **interior angles**.
  - ∠3 and ∠5 are **alternate interior angles** (opposite sides of the transversal).
  - Angles not between the parallel lines are **exterior angles**.
  - ∠1 and ∠7 are **alternate exterior angles** (opposite sides of the transversal).

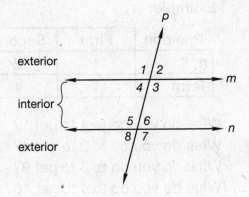

Practice:

Use the figure at right to answer questions 1–6.
Lines f and g are parallel.

1. Angle 3 measures 110°.

   What is the measure of ∠7? _____

2. Which angle is an alternate

   interior angle to ∠2? _____

3. Angle 4 measures 70°.

   What is the measure of ∠8? _____

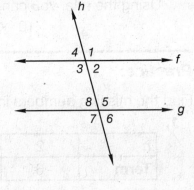

4. Which angle is an alternate exterior angle to ∠6? _____

5. Which line is a transversal? _____

6. Which angle corresponds to ∠1? _____

Name _____

Reteaching 98

Math Course 1, Lesson 98

• **Sum of the Angle Measures of Triangles and Quadrilaterals**

The sum of the interior angles of a triangle is 180°.

The sum of the interior angles of a quadrilateral is 360°.

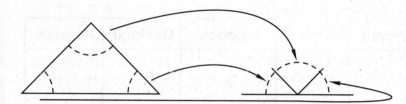

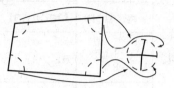

Examples:

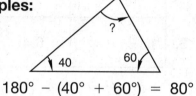

$180° - (40° + 60°) = 80°$

$360° - (120° + 80° + 80°) = 80°$

Practice:

1. Triangle *FGH* is an isosceles triangle. Angles *G* and *H* each measure 65°. What is the measure of angle *F*? _____

2. Triangle *ABC* is a right triangle. Angle *A* measures 90° and angle *B* measures 50°. What is the measure of angle *C*? _____

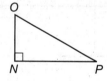

3. What is the measure of ∠*K* in quadrilateral *JKLM*? _____

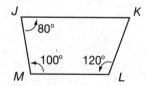

4. What is the sum of the measures of the four interior angles of a square? _____

Name _____

Reteaching 99

Math Course 1, Lesson 99

- **Fraction-Decimal-Percent Equivalents**

  - Fractions, decimals, and percents are three ways to express parts of a whole.
  - You can show equivalent fractions, decimals, and percents in a table.

	Fraction	Decimal	Percent
1.	$\frac{1}{2}$	a.	b.
2.	a.	0.3	b.
3.	a.	b.	40%

→

	Fraction	Decimal	Percent
1.	$\frac{1}{2}$	a. 0.5	b. 50%
2.	a. $\frac{3}{10}$	0.3	b. 30%
3.	a. $\frac{4}{10} = \frac{2}{5}$	b. 0.4	40%

Practice:

Complete this table.

	Fraction	Decimal	Percent
1.	$\frac{4}{5}$	b.	b.
2.	a.	a.	6%
3.	a.	1.7	b.

108 © Harcourt Achieve Inc. and Stephen Hake. All rights reserved. Saxon Math Course

Name _____

Reteaching 100
Math Course 1, Lesson 100

- **Algebraic Addition of Integers**

  - **Integers:** the set of numbers that includes all the counting numbers, their opposites, and zero
  - To add integers as illustrated on a number line:
    1. Begin at zero.
    2. Move right or left as indicated by the sign of the first number.
    3. Then move right or left as indicated by the second number.

    Example: +8 + −5

    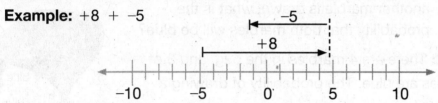

    The sum is +3 or 3.

  - To add signed numbers:
    If the signs are the same, add the absolute values and keep the same sign.
    If the signs are different, subtract the absolute values; keep the sign of the number with the greater absolute value.

    Examples: (−5) + (−3) = −8
    (+8) + (−5) = +3

  - Adding the opposite of a number to subtract is called **algebraic addition.**
    Instead of subtracting a negative, add a positive.
    Instead of subtracting a positive, add a negative.
    Change the number after a subtraction sign to its opposite and then add.

    Examples: −10 − (−6)   −3 − (+5)
    −10 + (+6) = −4   −3 + (−5) = −8

Practice:

Simplify 1–4.

1. −2 + −7 = _____

2. −3 − (−6) = _____

3. −9 − (−5) = _____

4. −4 + −8 = _____

5. At 6 a.m. the temperature was −4°F. By noon the temperature had risen to 10°F. How many degrees had the temperature risen? _____

6. At 6 a.m. the temperature was −9°C. By noon the temperature had risen to −1°C. How many degrees had the temperature risen? _____

Name _____

Reteaching Inv.

Math Course 1, Investigation 10

• **Compound Experiments**

Compound probability experiments have two or more parts.
To find the probability of a compound experiment:
1. Find the probability of each part.
2. Multiply the probabilities.

   Example: In a bag there are three blue marbles and one white marble. If one marble is drawn and not replaced and another marble is drawn, what is the probability that both marbles will be blue?

   First Part: There are 4 marbles in the bag, and 3 of the marbles are blue. The probability of drawing a blue marble is $\frac{3}{4}$.

   Second Part: Since a blue marble was chosen and not replaced on the first draw, there are now only 3 marbles in the bag, and 2 of the marbles are blue. The probability of drawing a blue marble this time is $\frac{2}{3}$.

   Multiply the probabilities. Then reduce your answer if you can. The probability that both marbles will be blue is $\frac{1}{2}$.

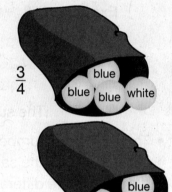

$$\frac{3}{4} \times \frac{2}{3} = \frac{6}{12} = \frac{1}{2}$$

Practice:

1. A bag contains 3 red marbles and 2 blue marbles. If one marble is drawn and not replaced and another marble is drawn, what is the probability that both marbles will be blue? _____

2. What is the probability of rolling a 2 twice with two rolls of a number cube? _____

3. A bag contains 5 red, 2 blue, and 3 yellow marbles. If one marble is drawn and not replaced and another marble is drawn, what is the probability that both marbles will be blue? _____

4. In a bag there are four white marbles and four blue marbles. If two marbles are drawn from the bag at the same time, what is the probability that both marbles will be white? _____

Reteaching 101

Math Course 1, Lesson 101

- **Ratio Problems Involving Totals**

In some **ratio** problems a **total** is needed in order to solve the problem.
- Make a table with the information in the problem.
- Include a row in the table for the total.
- Write a proportion.
- Use the row with what you want to find.
- Use the row that is complete.

Example: The ratio of boys to girls in a class was 5 to 4.
If there were 27 students in the class,
how many girls were there?

	Ratio	Actual Count
Boys	5	b
Girls	4	g
Total	9	27

$$\frac{4}{9} = \frac{g}{27}$$
$$9g = 4 \cdot 27$$
$$g = 12$$

Practice:

Draw a ratio box for each problem.
Then write and solve a proportion to find the answer.

1. The boy-girl ratio in the class was 3 to 5.

   If there were 24 students, how many boys were there? _____

2. The ratio of boys to girls in the ski club was 5 to 4.

   If there were 36 students, how many girls were there? _____

3. What is the boy-girl ratio in a class

   of 28 pupils if there are 12 girls? _____

4. What is the boy-girl ratio in a class

   of 32 pupils if there are 12 boys? _____

5. What is the boy-girl ratio on a team

   of 20 players if there are 8 boys? _____

Name _____

Reteaching 102

Math Course 1, Lesson 102

• Mass and Weight

- Physical objects are composed of **matter.**
- The amount of matter in an object is its **mass.**
- **Mass** does not change with changes in gravity.
- **Weight** does change with changes in gravity.
  The *weight* of an astronaut changes on the Moon.
  His or her *mass* does not change on the Moon.

Weight	Mass
U. S. Customary System	Metric System
16 oz = 1 lb 2000 lb = 1 ton	1000 g = 1 kg

Practice:

1. Three tons is how many pounds? _____

2. Three kilograms is how many grams? _____

3. Two pounds is how many ounces? _____

4. Half of a pound is how many ounces? _____

5. The mass of a liter of water is 1 kilogram.

   So, the mass of half of a liter of water is how many grams? _____

6. Half of a ton is how many pounds? _____

Name _____

Reteaching 103

Math Course 1, Lesson 103

• **Perimeter of Complex Shapes**

Perimeter means to add **all** the sides.
• Some sides will not be labeled.
• Add or subtract as needed to find the length of those sides.
• *Hint:* Sometimes it helps to use two different colors.
  Trace over all horizontal lines in one color.
  Trace over all vertical lines in another color.

Example:

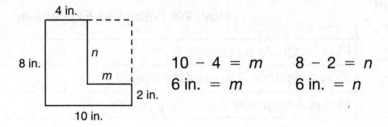

Add the lengths of all the sides to find the perimeter.

8 in. + 4 in. + 6 in. + 6 in. + 2 in. + 10 in. = 36 in.

Practice:

1. What is the perimeter of this figure?

   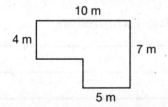

2. What is the perimeter of this figure?

   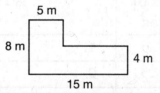

_____ _____

3. What is the perimeter of the hexagon?

   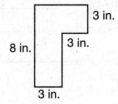

4. What is the perimeter of the hexagon?

   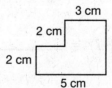

_____ _____

Name _____

Reteaching 104

Math Course 1, Lesson 104

• Algebraic Addition Activity

- Numbers greater than zero are written with a positive sign (+), or no sign at all. Numbers less than zero are always written with a negative sign (−).
- When an algebraic expression represents the addition or subtratction of positive and negative numbers, we look at the sign of the number to determine the mathematical operation we use to simplify the expression.
- Sometimes we enclose numbers in parentheses so that the sign of the number (negative sign) can be expressed separately from the operation (minus symbol).

Examples of Positives	How We Read the Expression
+ 2 = +2	Plus 2 equals a positive 2.
+ +2 = +2	Plus a positive 2 equals a positive 2.
− −2 = +2	Minus a negative 2 equals a positive 2.
−(−2) = +2	The negative of a negative 2 equals a positive 2.

Practice:

For each of the following, write the example of a negative number as you would read the expression.

Examples of Negatives	How We Read the Expression
− 2 = −2	1.
− +2 = −2	2.
+ −2 = −2	3.
−(+2) = −2	4.

Simplify 5–10.

5. (−5) + (−3) = _____

6. (−2) + (+6) = _____

7. (+1) + (−7) = _____

8. (9) + (−3) = _____

9. (+7) + (+6) + (−1) = _____

10. (−2) + (−9) + (11) = _____

© Harcourt Achieve Inc. and Stephen Hake. All rights reserved.

Saxon Math Course 1

Name _____

Reteaching 105

Math Course 1, Lesson 105

- **Using Proportions to Solve Percent Problems**

  - A **ratio box** may be used to solve percent problems. Use three rows. The total is 100%
  - Write a proportion using the complete row and the row with the information you want to find out.
  - Cross multiply and divide. Reduce when possible.

Example: Thirty percent of the students earned an A on the test. If twelve students earned an A, how many students were there in all?

	Ratio	Actual Count
A's	30	12
Not A's	70	n
Total	100	t

$$\frac{30}{100} = \frac{12}{t} \qquad t = \frac{\overset{4}{\cancel{12}} \cdot \overset{10}{\cancel{100}}}{\underset{1}{\cancel{\underset{3}{30}}}} = 40$$

Practice:

1. Sixty percent of the students who took the test earned an A. If twelve students earned an A, then how many students took the test?

   (Use a ratio box.) _____

2. Eighty percent of the students who took the test earned an A. If twenty students earned an A, then how many students took the test?

   (Use a ratio box.) _____

3. Leah missed 4 questions on the test but answered 80% of the questions correctly. How many questions were on the test?

   _____

4. Marco missed 6 questions on the test but answered 75% of the questions correctly. How many questions were on the test?

   _____

5. Ninety percent of the team members played in the game. If 18 members played, how many team members did not play?

   _____

Saxon Math Course 1

Name _____

Reteaching 106

Math Course 1, Lesson 106

- **Two-Step Equations**

Solve: $3n - 1 = 20$

1. When 1 is subtracted from $3n$, the result is 20. So $3n$ equals 21.
$$3n = 21$$

2. Since $3n$ means "3 times n" and $3n$ equals 21, we know that n equals 7.
$$n = 7$$

Check: $3(7) - 1 = 20$
$21 - 1 = 20$
$20 = 20$

Practice:

Solve 1–6.

1. $4n + 2 = 30$

   $n =$ _____

2. $3m - 3 = 21$

   $m =$ _____

3. $5y + 4 = 34$

   $y =$ _____

4. $7w + 3 = 17$

   $w =$ _____

5. $6k - 2 = 16$

   $k =$ _____

6. $3z + 1 = 28$

   $z =$ _____

Name _____

Reteaching 107
Math Course 1, Lesson 107

- **Area of Complex Shapes**

  - **Perimeter** of complex shapes ⟶ Add all sides.
  - **Area** of complex shapes
    1. Divide the shape into two or more parts.
    2. Find the area of each part.
    3. Add the parts.
    4. Sometimes it is easier to make a bigger rectangle and subtract a small part to find the area.
  - Formulas to remember:
    Area of a **rectangle** $A = lw$
    Area of a **triangle** $A = \frac{bh}{2}$
    (Be sure to label area in **square** units.)

Example:

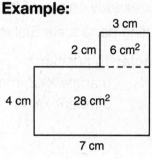

$$\begin{array}{r} 28 \text{ cm}^2 \\ + \ 6 \text{ cm}^2 \\ \hline 34 \text{ cm}^2 \end{array}$$

Practice:

1. What is the area of the hexagon?

2. What is the area of this figure?

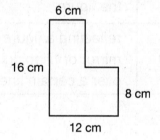

_____   _____

3. What is the area of this figure?

4. Find the area of this figure.

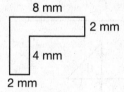   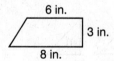

_____   _____

Name _____

Reteaching 108

Math Course 1, Lesson 108

- **Transformations**

  - Figures that have the same shape and size are **congruent**.
  - One will fit exactly on top of the other.
  - The matching parts are equal in measure.

    Example: To position triangle *ABC* on triangle *XYZ*, make three different kinds of moves.

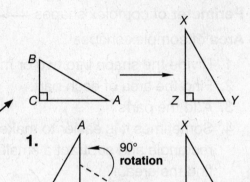

Transformations	
Rotation	turning a figure about a certain point
Translation	sliding a figure in one direction without turning the figure
Reflection	reflecting a figure as in a mirror or "flipping" a figure over a certain line

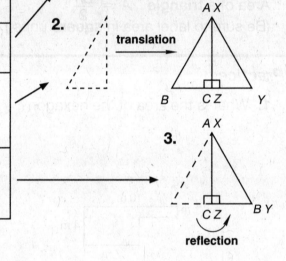

Practice:

Name the transformation(s) necessary to position triangle I on triangle II.

1. _____

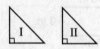

2. _____

3. _____

4. _____

Reteaching 109

Math Course 1, Lesson 109

- **Corresponding Parts**
- **Similar Triangles**

- If two figures are **congruent**, their corresponding parts (angles and sides) match exactly.

  Example: Triangle ABC and triangle XYZ are congruent.
  $\angle A$ corresponds to $\angle X$.
  $\overline{AB}$ corresponds to $\overline{XY}$.

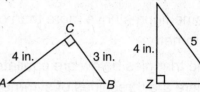

- Measures of corresponding parts of congruent figures are equal.

  Example: Triangle *ABC* and triangle *XYZ* are congruent. What is the perimeter of each?
  Since side *AB* corresponds to side *XY*, the length of side *AB* is 5 in. So, each triangle has sides that measure 3 inches, 4 inches, and 5 inches.
  4 in. + 3 in. + 5 in. = 12 in.
  The perimeter of each triangle is 12 inches.

- If two figures are **similar**, they have the same shape but not necessarily the same size. Similar figures have *equal matching* angles.

  Triangles I, II, and III are **similar**.
  Triangles I and II are **congruent**.

Practice:

1. The two triangles at right are congruent. What is the perimeter of each?

   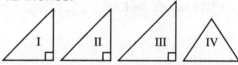

   _____

2. Which rectangles appear to be similar?

   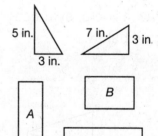

   _____

3. The two triangles at right are congruent. Which angle corresponds to $\angle F$?

   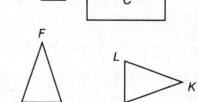

   _____

4. If two squares are congruent and the perimeter of one is 24 mm, what is the area of the other?

   _____

Name _____

Reteaching 11

Math Course 1, Lesson 110

- **Symmetry**

  - A figure is **symmetrical** if it can be divided in half so that the halves are mirror images of each other.
  - The line that divides a figure into two mirror images is called a **line of symmetry.**
  - Some figures have more than one line of symmetry. Some figures have no line of symmetry.

    The triangles below are equilateral. Each side is the same length.
    There are three lines of symmetry.

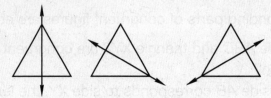

Practice:

1. Draw a line of symmetry on the figure below.

2. Draw a line of symmetry on the figure below.

   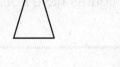

3. Draw a line of symmetry on the figure below.

4. Draw a different line of symmetry on each rectangle.

          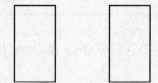

120 © Harcourt Achieve Inc. and Stephen Hake. All rights reserved.   Saxon Math Course 1

Reteaching Inv. 11

Math Course 1, Investigation 1

Name _____

Scale Drawings and Models

- **Scale drawings** are two-dimensional representations of larger objects.
- The **legend** gives the relationship between the unit of length in the drawing and the actual measurement that the unit represents.

The grid below is made of $\frac{1}{4}$-inch squares.

Actual Size

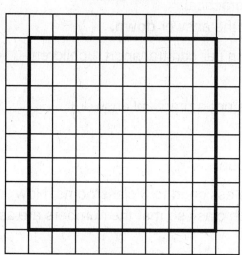

Length of 4 squares in grid = 1 inch

Scale Drawing
$\frac{1}{4}$ inch · 1 inch

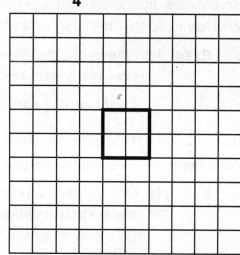

Length of 1 square in grid = 1 inch

Practice:

1. Draw the figure shown using the scale.

Actual Size

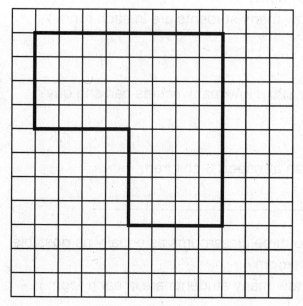

Scale Drawing
$\frac{1}{4}$ inch · 1 inch

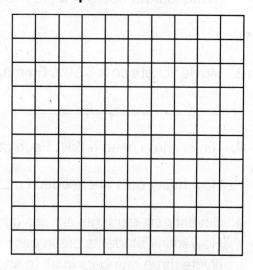

Name _____

Reteaching 11

Math Course 1, Lesson 111

- **Applications Using Division**

  - When a division problem has a remainder, there are several ways to write the answer.

with a remainder	as a mixed number	as a decimal number
3 R 3 4)15	$3\frac{3}{4}$ 4)15	3.75 4)15.00

  - Sometimes an answer with a remainder is not practical. We may need to round the answer **up** or round the answer **down**.

    Example: Fifteen children need a ride. Each car can transport 4 children. How many cars are needed?

    They need $3\frac{3}{4}$ cars. ($\frac{3}{4}$ of a car is not a practical answer.)

    We round $3\frac{3}{4}$ cars **up** to **4 cars.**

  - We may need to assign the remainders to a group.

    Example: One hundred students are to be assigned to 3 classrooms. How many students should be in each class so that the numbers are as balanced as possible?

    100 ÷ 3 = 33 R 1
    33 students, 33 students, and 34 students

Practice:

1. Ninety-seven students are assigned to four classrooms as equally as possible. How many students are in each classroom?
   (Write four numbers in all to show how many students are in each room.)

   _____

2. Movie tickets cost $7.00. Emma has $20. How many tickets can she buy?

   _____

3. Ten children need a ride. Each car can transport 3 children.

   How many cars are needed? _____

4. Seventy-six students are assigned to three classrooms as equally as possible. How many students are in each classroom?
   (Write three numbers in all to show how many students are in each room.)

   _____

Name _____

Reteaching 112
Math Course 1, Lesson 112

• **Multiplying and Dividing Integers**

To multiply or divide two integers:
1. Multiply or divide as indicated.
2. Place a sign with the answer.

If the signs of the two numbers are the same, the answer is positive.
  Positive × positive = positive   Negative × negative = positive
  Positive ÷ positive = positive   Negative ÷ negative = positive

If the signs of the two numbers are different, the answer is negative.
  Positive × negative = negative   Negative × positive = negative
  Positive ÷ negative = negative   Negative ÷ positive = negative

---

Practice:

Simplify 1–6.

1. $(-3)(-8)$ = _____

2. $(-6)(+8)$ = _____

3. $(7)(-5)$ = _____

4. $\dfrac{-27}{+9}$ = _____

5. $\dfrac{28}{-4}$ = _____

6. $\dfrac{-15}{-3}$ = _____

Name _____

Reteaching 11
Math Course 1, Lesson 113

- **Adding and Subtracting Mixed Measures**
- **Multiplying by Powers of Ten**

- To add or subtract **mixed measures,** units may need to be renamed.

    Example: 6 ft 5 in. $\overset{5}{\cancel{6}}$ ft $\overset{17}{\cancel{5}}$ in.
                  − 4 ft 8 in.   − 4 ft 8 in.
                                    ─────────
                                      1 ft 9 in.

- To multiply by powers of ten:
  Move the decimal point to the right the same number of places as the exponent.

    Examples: Write 1.2×10^3 in standard notation. $1.2 \times 10^3 = 1.200 = 1200$

    Write 6.2×10^2 in standard notation. $6.2 \times 10^2 = 6.20 = 620$

- Sometimes powers of ten are written as words instead of numbers.

    5.2 million means $5.2 \times 1{,}000{,}000$

- To write a number in standard notation:
    1. If a fraction is given, change it to a decimal.
    2. Write the power-of-ten word in number form and count the zeros.
    3. Move the decimal point right the same number of places as zeros.

    Example: Write $\frac{1}{2}$ million in standard notation.

      1. $\frac{1}{2} = 0.5$
      2. $0.5 \times 1{,}000{,}000 = 500{,}000$
      3. $0.500000. = 500{,}000$

Practice:

1. Write 3.5 million as a standard numeral. _____

2. Write $1\frac{1}{2}$ million as a standard numeral. _____

3. Write 2.7×10^3 in standard notation. _____

4. 5 ft 3 in.   5. 8 ft 2 in.
   − 3 ft 9 in.   − 2 ft 6 in.

6. 2 min 40 sec
   + 3 min 50 sec

Name _____

Reteaching 114
Math Course 1, Lesson 114

- **Unit Multipliers**

  - A **unit multiplier** is a fraction that equals 1 and is written with two different units of measure.

    3 feet = 1 yard $\frac{3 \text{ ft}}{1 \text{ yd}}$ and $\frac{1 \text{ yd}}{3 \text{ ft}}$

    Each fraction equals 1 because the numerator and the denominator of each fraction are equal.

    The units we change **from** are in the denominator.
    The units we change **to** are in the numerator.

    Example: Convert 30 feet to yards using a unit multiplier. (1 yd = 3 ft)

    $$\overset{10}{\cancel{30} \text{ ft}} \times \frac{1 \text{ yd}}{\cancel{3} \text{ ft}} = 10 \text{ yd}$$

Practice:

1. Write two unit multipliers for the equivalent measures 1 ft = 12 in.

   _____

2. Write two unit multipliers for the equivalent measures 1 pt = 2 c.

   _____

3. Show how to convert 5 feet to inches using a unit multiplier. (1 ft = 12 in.)

   _____

4. Show how to convert 10 cups to pints using a unit multiplier. (1 pt = 2 c)

   _____

5. Show how to convert 6 quarts to pints using a unit multiplier. (1 qt = 2 pt)

   _____

6. Show how to convert 12 quarts to gallons using a unit multiplier. (1 gal = 4 qt)

   _____

Name _____

Reteaching 115

Math Course 1, Lesson 115

• **Writing Percents as Fractions, Part 2**

Remember: a percent is a fraction with a denominator of 100.

To write a percent as a fraction:
• Remove the percent sign, write the number with a denominator of 100, and reduce the fraction.

Example: $50\% = \frac{50}{100} = \frac{1}{2}$

• If the percent includes a fraction, **divide** by 100.

Example: Convert $3\frac{1}{3}\%$ to a fraction. $\quad \frac{3\frac{1}{3}}{1} \div \frac{100}{1}$

$$\frac{\cancel{10}^{1}}{3} \times \frac{1}{\cancel{100}_{10}} = \frac{1}{30}$$

Practice:

1. Convert $8\frac{1}{3}\%$ to a fraction. _____

2. Convert $4\frac{1}{6}\%$ to a fraction. _____

3. Convert $7\frac{1}{7}\%$ to a fraction. _____

4. Convert $33\frac{2}{6}\%$ to a fraction. _____

5. Convert $53\frac{1}{3}\%$ to a fraction. _____

6. Convert $73\frac{1}{3}\%$ to a fraction. _____

Name _____

Reteaching 116

Math Course 1, Lesson 116

- **Simple Interest**

  - The amount of money deposited in a bank is the **principal.**
  - The bank offers to pay **interest,** a percentage of the money deposited.
  - Simple interest is paid on the principal only.

    Example: John deposited $100 in an account that pays 6% simple interest each year. After three years, how much money will be in the account?

$100.00	principal
$6.00	first-year interest
$6.00	second-year interest
+ $6.00	third-year interest
$118.00	total

Practice:

1. Emily deposited $200 in an account that pays 4% simple interest.

   After three years, how much money will be in the account? _____

2. Josh deposited $300 in an account that pays 3% simple interest.

   After three years, how much money will be in the account? _____

3. Zack deposited $400 in an account that pays 5% simple interest.

   After five years, how much money will be in the account? _____

4. Maria deposited $500 in an account that pays 7% simple interest.

   After four years, how much money will be in the account? _____

Saxon Math Course 1 © Harcourt Achieve Inc. and Stephen Hake. All rights reserved.

Name _____

Reteaching 117
Math Course 1, Lesson 117

- **Finding a Whole When a Fraction is Known**

To find a whole when a fraction is known:
1. Draw a picture divided into parts (denominator).
2. Mark off the parts (numerator).
3. Divide to find the number in each part. Label each part.
4. Find the total of the parts.

Example: $\frac{3}{8}$ of the people in the town voted. If 120 of the people voted, how many people lived in the town?
 1. 8 parts
 2. $\frac{3}{8}$ voted
 3. $120 \div 3 = 40$ in each part
 4. $8 \times 40 = 320$ people

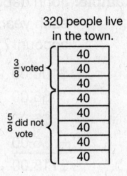

Practice:

1. Four fifths of the students in the class are boys. If there are 20 boys in the class, how many students are in the class?

   _____

2. Five eights of the players on the team are 12-years-old. If 10 of the players are 12-years-old, how many players are on the team?

   _____

3. Two thirds of the people in the pool are swimming in deep water. If there are 18 people in the deep water, how many people are in the pool?

   _____

4. Three fourths of the students in the class bought lunch. If 24 students bought lunch, how many students are in the class?

   _____

5. Six is $\frac{3}{5}$ of what number? _____

6. Fifteen is $\frac{3}{8}$ of what number? _____

Name _____

Reteaching 118

Math Course 1, Lesson 118

• **Estimating Area**

- To **estimate** the **area** of an irregular shape using a **grid:**
- **Shade** the squares that have **most** of their area within the shape.
- Put a **dot** in the squares that have about **half** of their area in the shape.
- Add the **shaded** squares plus **half** of the **dotted** squares.

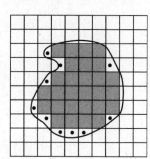

28 shaded squares
10 dotted squares
$28 + \frac{1}{2}(10) = 33$ square units

Practice:

1. Estimate the area of the irregular shape below.

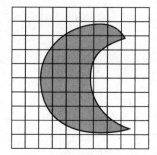

2. Estimate the area of the irregular shape below.

3. Estimate the area of the irregular shape below.

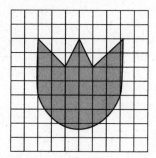

4. Estimate the area of the irregular shape below.

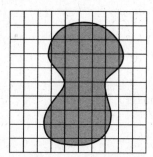

Name _____

Reteaching 11

Math Course 1, Lesson 119

• **Finding a Whole When a Percent Is Known**

To find a whole when a percent is known:
1. Translate the question into an equation.
2. Change the percent to a fraction or a decimal.
3. Solve.

Example: Thirty percent of what number is 120?

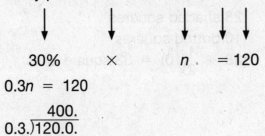

$0.3n = 120$

$$0.3 \overline{)120.0} = 400.$$

Thirty percent of **400** is 120.

Practice:

1. Twenty percent of what number is 100? _____

2. Fifty percent of what number is 375? _____

3. Forty percent of what number is 50? _____

4. Eighty percent of what number is 60? _____

5. Seventy-five percent of what number is 150? _____

6. Sixty-five percent of what number is 195? _____

Name _____

Reteaching 120

Math Course 1, Lesson 120

• Volume of a Cylinder

To calculate the volume of a cylinder:
1. Find the area of a circular end of the cylinder.
   Area = πr^2

Reminder: $\pi = 3.14$
$r = \frac{1}{2}d$

Example: $r = \frac{1}{2}(8 \text{ cm}) = 4 \text{ cm}$

$A = (3.14)(16 \text{ cm}^2)$

$A = 50.24 \text{ cm}^2$

$V = 50.24 \text{ cm}^2 \cdot 12 \text{ cm}$

2. Multiply that area by the height of the cylinder.
   Volume = $\pi r^2 \cdot$ height

$V = 602.88 \text{ cm}^3$

3. Round to the nearest whole number.

$V = 603 \text{ cm}^3$

Practice:

Round your answers to the nearest cubic centimeter.

1. What is its volume of this cylinder?

   _____

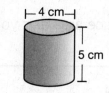

2. What is its volume of this cylinder?

   _____

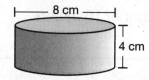

3. The diameter of a cylinder is 10 cm. Its height is 10 cm.

   What is its volume? _____

4. The diameter of a can is 12 cm. Its height is 15 cm.

   What is its volume? _____

5. The diameter of a cylinder is 20 cm. Its height is 15 cm.

   What is its volume? _____

6. The diameter of a can is 6 cm. Its height is 8 cm.

   What is its volume? _____

Saxon Math Course 1

Name _____

Reteaching Inv. 1

Math Course 1, Investigation 12

- **Volume of Prisms and Pyramids**
- **Surface Area of Prisms and Pyramids**

- A **prism** is a three-dimensional object (solid) with two congruent, parallel bases. A **pyramid** is a three-dimensional object (solid) that has three or more triangular **faces** and a base that is a polygon.

- We can use the **formula** for the volume of a prism to derive the formula for the volume of a pyramid. We find that we can "fit" exactly 6 similar pyramids inside a prism with the same base and twice the height of each pyramid.

    $V = B \times h$ Volume of a prism

    $V = \dfrac{B \times 2h}{6}$ Transform equation

    $V = \dfrac{1}{3}(B \times h)$ Volume of a pyramid

- The **surface area** of a prism is equal to the sum of the areas of its surfaces. The surface area of a pyramid is the sum of the areas of its base and its sides.

Practice:

1. Find the surface area and volume of a cube with a side length of 11 cm.

   _____

2. Find the volume of a pyramid with a base measuring 3 meters × 3.5 meters, and a height of 10 meters.

   _____

3. Find the surface area of a pyramid with a square base that has a side length of 10 mm and a slant height of 8 mm.

   _____

4. A pyramid with a volume of 32 cubic yards is inscribed in a cube with the same base area and height as the pyramid. What is the volume of the cube?

   _____

132   © Harcourt Achieve Inc. and Stephen Hake. All rights reserved.   Saxon Math Course 1